Applied Biocatalysis in Specialty Chemicals and Pharmaceuticals

ACS SYMPOSIUM SERIES **776**

Applied Biocatalysis in Specialty Chemicals and Pharmaceuticals

Badal C. Saha, EDITOR
Agricultural Research Service
United States Department of Agriculture

David C. Demirjian, EDITOR
ThermoGen, Inc.

American Chemical Society, Washington, DC

Chemistry Library

Library of Congress Cataloging-in-Publication Data

Applied biocatalysis in specialty chemicals and pharmaceuticals / Badal C. Saha, David C. Demirjian, editors.

p. cm.—(ACS symposium series ; 776)

".Developed from a symposium sponsored by the Division of Biochemical Technology, 217th ACS National Meeting, Anaheim, CA, March 21–25, 1999"

Includes bibliographical references and index.

ISBN 0–8412–3679–8

1. Enzymes—Biotechnology—Congresses. 2. Organic compounds—Synthesis— Congresses. 3. Pharmaceutical biotechnology—Congresses.

I. Saha, Badal C., 1949– II. Demirjian, David Charles, 1961– III. American Chemical Society. Division of Biochemical Technology (217th : 1999 : Anaheim, Calif.) IV.Series.

TD248.65.E59 A675   2000
660.6´34—dc21   00–63963

Foreword

The ACS Symposium Series was first published in 1974 to provide a mechanism for publishing symposia quickly in book form. The purpose of the series is to publish timely, comprehensive books developed from ACS sponsored symposia based on current scientific research. Occasionally, books are developed from symposia sponsored by other organizations when the topic is of keen interest to the chemistry audience.

Before agreeing to publish a book, the proposed table of contents is reviewed for appropriate and comprehensive coverage and for interest to the audience. Some papers may be excluded in order to better focus the book; others may be added to provide comprehensiveness. When appropriate, overview or introductory chapters are added. Drafts of chapters are peer-reviewed prior to final acceptance or rejection, and manuscripts are prepared in camera-ready format.

As a rule, only original research papers and original review papers are included in the volumes. Verbatim reproductions of previously published papers are not accepted.

ACS Books Department

Contents

Applications: Specialty Chemicals

Applications: Pharmaceutical

Preface

Tremendous advances have been made in the use of biocatalysts for the development of environmentally benign products and processes that can compete with conventional chemical processing. It is timely to organize a symposium on the subject and to provide a book that can assist practicing scientists and engineers with effective tools for tackling the future challenges in applied biocatalysts.

This volume was developed from a symposium presented at the 217th National Meeting of the American Chemical Society (ACS) entitled "Advances in Applied Biocatalysis," sponsored by the Division of Biochemical Technology, in Anaheim, California, March 21–25, 1999. This book presents a compilation of eleven of those manuscripts and five solicited manuscripts representing recent advances in the development and application of biocatalysis in the production of fine chemicals and pharmaceuticals. The chapters in the book have been organized in three sections. The first section, "Biocatalyst Discovery, Characterization, and Engineering," describes new technologies and advances that are being used to rapidly discover and create new enzymes and biocatalytic abilities. The second section, "Applications: Specialty Chemicals," is devoted to describing the use of biocatalysts in chemical processes. Finally, "Applications: Pharmaceutical," describes fine chemical applications such as the synthesis of chiral molecules and pharmaceutical intermediates. An overview chapter on applied biocatalysis has been included.

We are fortunate to have contributions from world-class researchers in the field of biocatalysis. We take this opportunity to express our sincere appreciation to the contributing authors, the reviewers who provided excellent comments to the editors, the ACS Division of Biochemical Technology, and the ACS Books Department for making possible the symposium and the publication of the book.

We hope that this book will actively serve as a valuable multidisciplinary contribution to the continually expanding field of biocatalysis.

BADAL C. SAHA
Termination Biochemistry Research Unit
National Center for Agricultural Utilization Research
Agricultural Research Service
U.S. Department of Agriculture
1815 North University Street
Peoria, IL 61604

DAVID C. DEMIRJIAN
ThermoGen, Inc.
2225 West Harrison
Chicago, IL 60612

Overview

Chapter 1

Advances in Enzyme Development and Applied Industrial Biocatalysis

Badal C. Saha[1] and David C. Demirjian[2]

[1]Fermentation Biochemistry Research Unit, National Center for Agricultural Utilization Research, Agricultural Research Service, U.S. Department of Agriculture, 1815 North University Street, Peoria, IL 61604
[2]ThermoGen Inc., 2225 West Harrison, Chicago, IL 60612

Biocatalysts play important roles in various biotechnology products and processes in the food and beverage industries and have already been recognized as valuable catalysts for various organic transformations and production of fine chemicals and pharmaceuticals. At present, the most commonly used biocatalysts in biotechnology are hydrolytic enzymes which catalyze the breakdown of larger biopolymers into smaller units. Enzymes catalyze reactions in a selective manner, not only regio- but also stereoselectively and have been used both for asymmetric synthesis and racemic resolutions. The chiral selectivity of enzymes has been employed to prepare enantiomerically pure pharmaceuticals, agrochemicals and food additives. Biocatalytic methods have already replaced some conventional chemical processes. Biocatalytic routes, in combination with chemical synthesis, are finding increased use in the synthesis of novel polymeric materials. The present global market for enzymes is estimated to be more than US $1.5 billion. The discovery of new and improved enzymes and their use in various processes and products will create new market opportunities for biocatalysts and helps solve environmental problems.

Applied biocatalysis can be defined as the application of biocatalysts to achieve a desired conversion under controlled conditions in a bioreactor (*1*). A biocatalyst can be an enzyme, an enzyme complex, a cell organelle or whole cells. The source of biocatalyst can be of microbial, plant or animal origin. Catalysis by an enzyme offers a number of advantages over traditional chemical catalysis. Enzymes as biocatalysts are both efficient with high catalytic power and highly specific for a particular chemical reaction involving the synthetic, degradative or alteration of a compound. They increase the rate of chemical reaction by factors 10^9 to 10^{12}. Enzymes work under mild conditions of temperature, pH and pressure. They are also highly biodegradable and generally pose no threat to the environment.

Enzymes, produced by living systems, are proteineous in nature. Cofactors are involved in reactions where molecules are oxidized, reduced, rearranged or connected. Enzymes have been divided into six major classes based on the types of reactions they catalyze:

1. Hydrolases: catalyze hydrolytic reactions (glycosidases, peptidases, esterases). Water is the acceptor of the transferred group.
2. Oxido-reductases: catalyze oxidation or reduction reactions (dehydrogenases, oxidases, peroxidases).
3. Isomerases: catalyze isomerization and racemization reactions (racemases, epimerases).
4. Transferases: catalyze the transfer of a group from one molecule to another one (glycosyl transferases, acetyl transferases).
5. Lyases: catalyze elimination reactions where a bond is broken without oxidoreduction or hydrolysis (decarboxylases, hydrolyases).
6. Ligases: catalyze the joining of two molecules with ATP or other nucleoside triphosphate cleavage (DNA ligases).

Microbial enzymes have largely replaced the traditional plant and animal enzymes used in industry. At present, about 50 enzymes are used in industry, most of them ($\sim$ 90%) are produced by submerged or solid state fermentation by microorganisms. Most industrial enzymes are produced extracellularly. Major exceptions are glucose isomerase, invertase and penicillin acylase. This chapter provides an overview of biocatalysis from discovery to applications in the food, pharmaceutical, chemical and medical diagnostic industries and the future of biocatalysis in these fields.

Discovery and Engineering of Biocatalysts

One of the largest impediments to the development of biocatalytic processes in the past was the discovery and engineering of biocatalysts for specific commercial applications. Researchers trying to develop bioprocesses were limited to a relatively small number of enzymes that had been previously discovered or studied, often for entirely different applications. In addition, only enzyme properties and preferences that had naturally evolved were available - unless one used extremely expensive and time-consuming rational engineering approaches. These approaches often had unpredictable results and could only be used on enzymes that were extremely well characterized. This made the development of bioconversion processes for unnatural substrates (such as pharmaceuticals) very difficult.

Today, the discovery and engineering of novel biocatalysts is becoming increasingly more attainable. New enzyme properties can be found by either screening from natural sources or gene libraries, or by creating novel activities through directed evolution.

Enzyme Discovery

Enzyme Sources

A number of sources are now available for the researcher who wishes to develop a biocatalytic process. The fastest and easiest route is to find an enzyme from a commercial library. The biggest change over the last several years has been the development of larger commercial enzyme libraries to enhance and simplify biocatalyst discovery. Traditional sources such as Novo Nordisk, Sigma, Amano, Roche Molecular Biochemicals and Toyobo have been joined by new companies such as ThermoGen and Diversa which offer an expanded range of enzymes which can be adapted to biocatalytic processes.

Still, it is not always possible to find an appropriate enzyme from a commercial source, requiring a custom screening effort. Screening can be carried out from a collection of microorganisms or from clone banks that have been generated from these organisms or isolated DNA.

Native Strain Sources

Screening from culture sources has been the historical method of finding new enzymes and has been successful in many cases (2). Most enzymes of industrial importance developed in the past have been derived from species that fall under the GRAS classification (Generally Regarded as Safe). These include bacterial species for *Bacillus* and *Lactobacillus*, and *Pseudomonas* and fungi from the *Ascomycota* and *Zygomycota* classes (3). If one knows which type of enzyme one is screening for, cultures can often be enriched for particular enzyme activities by standard methods (4).

Screening from a culture source can pose several challenges. Since the media for different organisms and protein expression conditions vary, the systematic screening of organism banks becomes more difficult. The cost of establishing and maintaining a proprietary strain collection can also be high. Strain redundancy is a concern, and verification that a particular strain is unique in a collection can be accomplished by several methods. Phenotypic (5), ribosome relationship (6) or PCR-based strain analysis (7) are all characterization methods useful in determining uniqueness.

Clone Banks and Expression Libraries

Screening for new enzymes from clone-banks can be rewarding. By setting up the clone banks in a unified or small set of host organisms (like *E. coli*, *Bacillus* or yeast), only a limited number of different propagation methods need to be implemented, thus allowing a systematic screening approach. When an enzyme is discovered from a clone bank, it is generally easier to scale-up and produce in larger quantities. In addition, genetic modification of the gene (such as directed evolution) is easier. The gene of interest may also be removed from its regulatory elements that can repress expression. The DNA used for cloning can originate from DNA prepared from cultured organisms or from uncultured organisms. It has been estimated that less than 1% of world's organisms have been cultured and techniques to isolate nucleic acids directly from soil samples can allow access to new genes (8). However, expressing DNA fragments from highly divergent organisms may be extremely difficult and identification of the organism class that the gene came from is difficult or impossible (this is important for GRAS applications).

There are also disadvantages to screening for new enzyme activities from clone banks due to removal of a gene from positive regulatory elements, host strain codon usage and nucleic acid structure or lethality issues. The activity from enzymes that are post-translationally modified may be altered or destroyed (9). In addition, for each organism a clone bank is developed from, one needs to screen thousands to tens of thousands of clones for each organism to cover the entire genome of that organism.

Enzyme Engineering

One of the technologies that has generated the most excitement recently is directed evolution. Through directed evolution, the properties of an enzyme can be fine-tuned by evolving it *in vitro* to enhance enzyme properties including activity (*10-12*). By introducing random mutations throughout an appropriate gene template, and screening or selecting for altered properties (such as pH optimum, reaction kinetics, solvent systems, enzyme expression and thermostability) improved enzymes can be identified. The process can be reiterated and mutations can often be combined to enhance activity further.

There are three basic steps in directed evolution where technology can be applied to enhance an enzyme by directed evolution. The first is the method that is used to generate a pool of mutants. PCR mutagenesis is often employed, and while the approach is not without its flaws, is a very reliable approach for generating random mutant libraries. The second area where technology can be applied is in the screen or selection that is used in finding mutants of interest. This is potentially the most important technology since it is required for finding the "needle in the haystack." Finally, methods for recombining mutations that are found can be used to combine advantageous mutations (*13, 14*).

Directed evolution is especially useful when combined with an enzyme discovery program. One can find an enzyme template that has activity close to the desired activity and use directed evolution to tweak it closer to the custom activity desired.

Screening Strategies

Whether screening for new enzymes from natural sources, mutant DNA pools or even screening enzyme properties, screening techniques are playing an increasingly important role in identifying enzymes of interest for applications. The availability of effective assays has been a major bottleneck in the development of a biotransformation-based process for industrial synthesis.

Screening Strategy

One of two methods is generally employed when carrying out a screening program on either clone banks or native strain collections – a brute force method or a hierarchical screening method. Each approach has its advantages. In the brute force method, each individual candidate is tested and analyzed for activity against a particular substrate. Libraries are generally arrayed in microtiter plates so that they can be systematically screened. Arraying all of the colonies can be extremely time consuming, but is often important if one cannot develop an appropriate random plate screen or selection to be used in a hierarchical screen.

One can gain significant increases in throughput by implementing a hierarchical screening approach, which combines several screening assays in a sequential fashion. Using this method the easier, but perhaps less accurate, screens are carried out first. The more tedious but quantitative screens are then carried out on only a fractional subset of candidates which have been pre-validated as potentially useful isolates. This type of screening approach is rapid, useful and cost effective, but accuracy requires the development of powerful assays.

Screening Substrates and Techniques

Occasionally, one can develop a genetic selection for enzyme activity which helps increase throughput enormously, but generally, one of a number of detection methods are employed in an enzymatic assay. It is most desirable to use the actual substrate of interest when screening for an enzyme activity, but unfortunately it is not always possible to use the actual substrate so an analog must be used. Generally these analogs are easy to assay and can, in some cases, resemble the substrate of interest. The use of substrate analogs almost assures that some potential candidates will be missed which act on a particular substrate of interest, or that some will be found that do not perform on the actual target substrate. For this reason, it is important to try and pick substrates that give a good cross-section of activities from the library being screened. For example, with hydrolases, a number of substrates are commonly used (*15*). One class of substrates is the precipitable indigogenic substrates. These are often ideal for first-level hierarchical plate assays since the color develops and stays in the vicinity of the colony. The other two classes of substrates are both soluble substrates that are useful in second-level screens since they are quantitative. Chromogenic substrates, such as those based on nitrophenyl or nitroaniline, can be used in a quantitative spectrophotometric liquid-assay. These are generally not useful in plate screens since they diffuse readily and are not sensitive enough. Fluorogenic substrates such as those based on umbelliferone or coumarin are at least 1,000 times more sensitive than their chromogenic counterparts.

As enzyme libraries and directed evolution applications grow and high throughput screening methods develop, new types of assays which utilize the specific target substrates instead of substrate analogs are needed. Some newer activity-based screening techniques are now being developed to get over the limitations of using substrate analogs. For example, a screening system to evaluate hydrolytic activities in liquid phase using pH indicators for mutants obtained by directed evolution techniques has been developed. Both qualitative and quantitative methods have been developed that can help rapidly assess enzyme activity (*16, 17*). This method is also useful for carrying out bioreaction engineering and optimization by allowing high throughput screening of different reaction conditions.

Application of Biocatalysts

The development of new methods for screening, the creation of enzyme libraries, the development of diverse organism and clone banks, the assaying of enzyme activity, and the evolution of proteins promise a breadth of new biocatalytic applications in the near future. However, a number of historical applications have been developed which demonstrate the power of enzymes and biocatalysis.

Detergents

The largest application of microbial enzymes is the use of proteases (pH 9-10, 50-60°C) in detergents. Lipase, amylase and cellulases are also used in addition to protease. Currently, the detergent industry occupies 25-30% of the entire industrial enzyme market. Enzyme containing detergents will continue to gain popularity. Cellulases are now used in biopolishing and stone-washing processes.

High Fructose Corn Syrups

The bioprocessing of starch to glucose and then glucose to fructose is a good example of the successful application of biocatalysis in an industrial scale. Three major enzymes are used: α-amylase, glucoamylase and glucose isomerase (GI). First, an aqueous slurry of starch (30-35% dry substance basis) is gelatinized (105°C, pH 6.0-6.5, 50 ppm Ca^{2+}, 5 min) and partially hydrolyzed (95°C, 2 hr) by a highly thermostable α-amylase to a DE of 10-15. Then the temperature is lowered to 55-60°C, pH is lowered to 4.0-5.0, and glucoamylase with or without pullulanase, a starch debranching enzyme, is added to continue the reaction for 24-72 hr, depending on the enzyme dose and the percent of glucose desired in the product. GI is produced intracellularly by *Streptomyces*, *Bacillus*, *Arthobacter* and *Actinoplanes*. High fructose corn syrups (HFCS) are prepared by enzymatic isomerization of glucose syrups (DE 95-98, 40-50% DS, pH 7.5-8.0, 55-60°C, 5 mM Mg^{2+}) in a column reactor containing immobilized GI. Mg^{2+} works as an activator and stabilizer of GI, which also compensates for the inhibitory action of Ca^{2+} on the enzyme. The use of immobilized GI allows a continuous process and avoids introduction of the enzyme into the process. GI is the biggest selling immobilized enzyme in the world.

6-Amino Penicillanic Acid

A variety of semisynthetic penicillin antibiotics is synthesized from 6-amino penicillanic acid (6-APA). The process for the production of 6-APA involves hydrolysis of penicillin G or V to form 6-APA followed by resynthesis using a different side chain. Chemical deacetylation to produce 6-APA was used originally, but the β-lactum ring is labile and the process requires the use of low temperatures, absolute anhydrous conditions and organic solvents, making the process difficult and expensive. Approximately 7500 tons of 6-APA are produced annually worldwide, mainly by deacylation of the native penicillins with immobilized penicillin amidase derived from *Escherichia coli* or *Bacillus megatetium* (*18*). This process is the best known use of an immobilized enzyme in the pharmaceutical industry (*1*).

Acrylamide

The conventional process for the manufacture of acrylamide involves copper-catalyzed hydration of the nitrile, which produces a number of toxic byproducts (*19*). Nitto Chemical Industry (Japan) uses nitrile hydratase from *Rhodococcus rhodochrous* J1 to convert acrylonitrile into acrylamide in a simple, clean and rapid process without the formation of unnecessary byproducts. The activity of the cells can be increased by the addition of ferric ions, by using methacrylamide as an inducer and by mutagenesis. The process has the advantage of allowing concentrated reactants to be used at low temperatures and of producing such a pure product that no purification is required prior to polymerization. *R. rhodochrous* J1 is also used in the industrial production of vitamin nicotinamide from 3-cyanopyridine. Nicotinamide is used as an animal feed supplement.

6-Hydroxynicotinic acid

6-Hydroxynicotinic acid (6-HNA) is a very useful intermediate in the synthesis of pesticides and pharmaceuticals as specific inhibitors of NAD and/or NADP dependent enzymes (*20*). *Achromobacter xylosoxydans, Pseudomonas acidovorans* and *P. putida* have been used to carry out the selective yield, using niacin hydroxylase which catalyzes the conversion of niacin to 6-HNA. The hydroxylation is oxygen requiring, so that oxygen transfer rate limits the reaction.

Lactose Hydrolysis

The enzyme lactase (ß-galactosidase) hydrolyzes lactose to glucose and galactose. People who suffer from lactase deficiency cannot drink milk which has 4.3-4.5% lactose. The enzyme is produced by many yeasts such as *Klyveromyces lactis* and fungi such as *Aspergillus niger*. However, it is inhibited by galactose. Lactase is still a relatively expensive enzyme (relative to the value-added to the substrate). Immobilized enzyme is used in industry to overcome the product inhibition problem and also to reduce enzyme treatment costs. The crystallization of lactose in ice cream products can be prevented by prior treatment of milk with immobilized lactase. Use of whey hydrolyzate as fermentation feedstock has been developed (*20*).

Indigo Dye

Indigo dye is used for dyeing of clothes, particularly denims. The manufacture of indigo dye requires à harsh chemical process and generates carcinogenes and toxic wastes. Amgen developed a biocatalytic production process for indigo dye. The pathway forming indigo involves converting tryptophan to indole via tryptophanase, then indole to cis-indole 2,3 glycol via napthalene dioxygenase, followed by non-enzymatic steps via indoxyl to indigo (20). The biotransformation process developed was not sufficiently efficient to easily compete with the traditional sources of indole.

Cocoa Butter

Cocoa butter is an important ingredient (~ 30%) in chocolate because of its unusual and useful melting behavior. The main triglycerides in cocoa butter are 1,3-distearoyl-2-oleoyl-glycerol and 1-stearoyl-2-oleoyl-3-palmitoyl-glycerol (20). Unilever has patented a process using a fixed bed reactor containing immobilized lipase to convert palm oil and stearic acid to a cocoa butter substitute (21).

De-bittering of Fruit Juices

Biocatalytic de-bittering of grapefruit juice can be achieved through the application of fungal naringinase preparations. The enzyme preparation contains both α-rhamnosidase (EC 3.2.1.40) and β-glucosidase activities. α-rhamnosidase first breaks down naringin [an extremely bitter flavanoid, 7-(2-rhamnosido--glucoside)] to rhamnose and prunin and then β-glucosidase hydrolyzes prunin to glucose and naringenin. Prunin bitterness is less than one third of that of naringin. However, α-rhamnosidase is competitively inhibited by rhamnose and β-glucosidase is inhibited by glucose. Immobilized enzymes are used to solve the inhibition problems. Another enzyme glucose oxidase (EC 1.1.3.4) is used to scavenge oxygen in fruit juice and beverages to prevent color and taste changes. Glucose oxidase is produced by various fungi such as *Aspergillus niger* and *Penicillium purpurogenum*. Pectic enzymes are used to increase fruit juice yield and to clarify juices (22).

Aspartame

Aspartame (L-aspartyl-L-phenylalanine methyl ester) is an artificial dipeptide sweetener about 200 times sweeter than sucrose and has a market of more than $1 billion. Aspartame can now be made by using an enzymatic process. The enzyme thermolysin (a metalloprotease produced by *B. thermoproteolyticus*) catalyzes the amide bond formation between L-aspartic acid and phenylalanine methyl ester. It is enantioselective and forms the peptide bond only with L-phenylalanine methyl ester. The enzyme is also regioselective and does not react with the β-carboxy of the aspartic acid, so no bitter tasting β-aspartame is formed. L-Aspartic acid can also be prepared enzymatically from ammonium fumarate substrate by a single enzyme L-aspartate ammonia lyase obtained from *Escherichia coli*.

Other Compounds

Ephedrine is widely used in the treatment of asthma and hay fever as a bronchodilating agent and decongestant. It is produced chemoenzymatically from benzaldehyde and pyruvate. At first, optically active phenyl-acetylcarbinol is produced from benzaldehyde and pyruvate by using brewers yeast and cell-free yeast extracts which is then reductively aminated to produce optically active L-ephedrine (*20*). Epoxide hydrolyases are ubiquitous enzymes able to hydrolyze an epoxide to its corresponding vicinal diol. These hydrolases have been shown often to be highly enantio- and regioselective, thus allowing both the epoxide and the diol to be prepared at high enantiomeric purity (*23*). A biocatalytic alternative to the usually employed industrial synthesis of catechol has been developed using glucose as substrate (*24*). *Klebsiella pneumoniae* genes encoding 3-dehydroshikimate dehydratase (*aroZ*) and protocatechuic acid decarboxylase (*aroY*) were introduced into an *Escherichia coli* construct that synthesizes elevated levels of 3-dehydroshikimic acid. One of the resulting biocatalysts synthesizes 18.5 ± 2.0 mM catechol from 56 mM glucose on 1L scale.

Analytical Applications of Biocatalysts

Enzymes are used in various analytical methods, both for medical and non-medical purposes (*25*). Immobilized enzymes, for example, are used as biosensors for the analysis of organic and inorganic compounds in biological fluids. Biosensors have three major components: a biological component (e.g., enzyme, whole cell), an interface (e.g., polymeric thick or thin film) and a transducing element which converts the biochemical interaction into a quantifiable electrical or optical signal. A glucose biosensor consists of a glucose oxidase membrane and an oxygen electrode while a biosensor for lactate consists of immobilized lactate oxidase and an oxygen electrode. The lactate sensor functions by monitoring the decrease in dissolved oxygen which results from the oxidation of lactate in the presence of lactate oxidase. The amperometric determination of pyruvate can be carried out with the pyruvate oxidase sensor, which consists of a pyruvate oxidase membrane and an oxygen electrode. For the determination of ethanol, the biochemical reaction cell using an alcohol dehydrogenase (ADH, EC 1.1.1.1) membrane anode is used.

A bioelectrochemical system for total cholesterol estimation was developed, based on a double-enzymatic method. In this system, an immobilized enzyme reactor containing cholesterol esterase (EC 3.1.1.13) and cholesterol oxidase (EC 1.1.3.6) is coupled with an amperometric detector system. An amino acid electrode for the determination of total amino acids has also been developed using the enzymes L-glutamate oxidase, L-lysine oxidase and tyrosinase. Enzyme electrodes are used for continuous control of fermentation processes.

Waste Treatment

Biocatalysts have great potential for degrading pollutants (26). Enzymes have been found that will detoxify organophosphate insecticides (27). The fungus *Stemphylium loti* degrades cyanide in waste streams efficiently (28).

Concluding Remarks

The development of new biocatalysts as synthetic tools has been expanding over the past several years. It is now possible either to discover or engineer enzymes with unique substrate specificities and selectivities that are stable and robust for organic synthetic applications. Advancements in the application of biocatalysts to industry has allowed faster development of biocatalytic processes, and new application areas have opened up over the last several years to take advantage of these advancements in new technologies.

References

1. Tramper, J. In *Applied Biocatalysis;* Cabral, J. M. S.; Best, D.; Boross, L.; Tramper, J., Eds.; Hardwood Academic Publishers, Switzerland, **1994**, pp 1-45.
2. Chartrain, M.; Armstrong, J.; Katz, S.; King, S.; Reddy, J.; Shi, Y. J.; Tschaen, D.; Greasham, R. *Ann. NY Acad. Sci.* **1996**, *799*, 612-19.
3. Dalboge, H. ; Lange, L. *TIBETCH* **1998**, *16*, 265-72.
4. Hummel, W. In *Frontiers in Biosensorics I: Fundamental Aspects,* Scheller, F. W.; Schubert, F.; Fedrowitz, J., Eds.; Birkhäuser Verlag, Basel, **1997**, pp 49-61.
5. Bergey, D. H.; Holt, J. G. In *Bergey's Manual of Determinative Bacteriology;* Williams & Wilkins, Baltimore, **1993**, cm.
6. Giovannoni, S. J.; DeLong, E. F.; Olsen, G. J.; Pace, N. R. *J. Bacteriol.* **1988**, *170*, 720-6.
7. Busse, H. J.; Denner, E. B. M. ; Lubitz, W. *J. Biotechnol.* **1996**, *47*, 3-38.
8. Tiedje, J. M.; Stein, J. L. In *Manual of Industrial Microbiology and Biotechnology;* Demain, A. L.; Davies, J. E.; Atlas, R. M., Eds.; ASM Press, Washington, DC, **1999**, pp 682-692.
9. Sambrook, J.; Maniatis, T.; Fritsch, E. F. *Molecular Cloning: A Laboratory Manual;* Cold Spring Harbor Laboratory, Cold Spring Harbor, NY, **1989**.
10. Schmidt-Dannert, C. ; Arnold, F. H. *TIBTECH* **1999**, *17*, 135-6.
11. Shao, Z. ; Arnold, F. H. *Curr. Opin. Struct. Biol.* **1996**, *6*, 513-8.
12. Arnold, F. H.; Volkov, A. A. *Curr. Opin. Chem. Biol.* **1999**, *3*, 54-9.
13. Stemmer, W. P. *Nature* **1994**, *370*, 389-91.
14. Zhao, H.; Giver, L.; Shao, Z.; Affholter, J. A.; Arnold, F. H. *Nat. Biotechnol.* **1998**, *16*, 258-261.

15. Miller, J. H. *A Short Course in Bacterial Genetics: A Laboratory Manual and Handbook for Escherichia coli and Related Bacteria;* Cold Spring Harbor Laboratories Press, Cold Spring Harbor, NY, **1991**.
16. Morís-Varas, F.; Shah, A.; Aikens, J.; Nadkarni, N. P.; Rozzell, J. D.; Demirjian, C. C. *Bioorg. Med. Chem.* **1999**, *7,* 2183-8.
17. Janes, L.; Löwendahl, C. ; Kazlauskas, R. Eur. J. *Chem.* **1998**, *4,* 2317.
18. Katchalski-Katzir, E. *TIBTECH* **1993**, *11*, 471-478.
19. Ogawa, J.; Shimizu, S. *TIBTECH* **1999**, *17*, 13-21.
20. Cheetham, P. J. In *Applied Biocatalysis;* Cabral, J. M. S.; Best, D.; Boross, L.; Tramper, J.; Eds.; Hardwood Academic Publishers, Switzerland, **1994**, 47-108.
21. Macrae, A. R. In *Biocatalysts in Organic Synthesis;* Tramper, J.; van der Plas; H. C.; Linko, P.; Eds.; Elsevier, Amsterdam, **1985**, pp 195-208.
22. Alkorta, I.; Garbisu, C.; Llama, M. J.; Serra, J. L. *Process Biochem.* **1998**, *33*, 21-28.
23. Archelas, A.; Furstoss, R. In *Biocatalysis from Discovery to Application*, Fessner, W.-D., Ed.; Springer-Verlag, Berlin Heidelberg, **1999**, pp 159-191.
24. Draths, K. M.; Frost, J. W. *J. Amer. Chem. Soc.* **1995**, *117*, 2395-2400.
25. Saha, B. C.; Bothast, R. J. In *Encyclopedia in Microbiology*, Academic Press, San Diego, CA, **2000**, Vol. 2, pp E92-E105.
26. Cheetam, P. S. J. *Enzyme Microb. Technol.* **1987**, *9*, 194-213.
27. Munnecke, D. M. *Appl. Environ. Microbiol.* **1976**, *32*, 7-13.
28. Nazaly, N.; Knowles, C. J. *Biotechnol. Lett.* **1981**, *3*, 363-368.

Biocatalyst Discovery, Characterization, and Engineering

Chapter 2

Selected Historical Perspectives in Biocatalysis, 1752–1960

S. L. Neidleman

Neidleman Consulting, 5377 Hilltop Crescent, Oakland, CA 94618

The purpose of this chapter is to offer a short glance back in the development of commercial applications of enzyme technology. It is not a matter of paying homage to the pioneers of biocatalysis, but rather an attempt to achieve a sense of historical perspective as an adjunct to understanding the present and anticipating the future. This goal will be approached by illustrating many examples of enzyme applications in the period 1752-1960, spiced by selected vignettes to inject humanity into this view of the past.

This is not to suggest that the use of enzymes to produce goods for human consumption began in the late 19th century. For example, the origins of biocatalysis in the dairy and brewing industries lie veiled in the mists of prehistory. Consider the amazement of the accidental inventor in pre-patent days who carried milk in the stomachs of animals and discovered cheese, or agitated milk in the heat of the midday sun and obtained butter, or the innocent taster of aged crushed fruit who grew tipsy with alcohol and elation. Not much has changed, although technology, commercialization and government regulations have civilized these and other similarly ancient biocatalytic processes.

WHAT IS CATALYSIS?

Early pronouncements on catalysis and then biocatalysis were a polyphonic mix of several themes, including: *(1)* what is a catalyst, *(2)* what is a biocatalyst and *(3)* what is the structure of a biocatalyst.

A survey of the more prominent visions will indicate the focus of the intellectual exertion exercised through a 100-year period.

In 1836, Berzelius defined a catalyst as a substance capable of wakening energies dormant at particular temperature, merely by its presence, Waksman and Davidson

(1); Fruton *(2)*. Further, Berzelius prophetically recognized the similarity of catalysis in the chemical laboratory and the living cell.

Such speculations did not, however, go unchallenged. Liebig stated that the assumption of this new force was detrimental to scientific progress, satisfying, rather, the human spirit. He suggested that any catalytic agent was itself unstable and during its decomposition caused unreactive substances to undergo chemical change.

The challenger was then challenged. In 1878, Traube retorted that ferments were not unstable substances transmitting chemical vibrations to unreactive materials, but were chemicals, related to proteins, possessing a definite chemical structure that evoked changes in other chemicals through specific chemical affinities, Fruton and Simmonds *(3)*.

In the 18th and 19th centuries, and even into the 20th century, a dominant concept was that referred to as "Lebenskraft" or "Spiritus Vitae." Among its tenets was the idea that minerals (inorganics) were products of ordinary physical forces, but organic compounds owed their synthesis to an organic vital force associated with the living cell. Even Berzelius in 1827 believed that organic synthesis in the laboratory was impossible. Organic chemists struggled with this belief for well over 100 years.

Levene *(4)* wrote that the story of the rise and fall of biochemistry in the esteem of the scientific community is connected to its revolt against the concept of the vital force. The biocatalysts were clearly enmeshed in this controversy. There was the distinction between formed or organized ferments such as yeast and the unorganized ferments such as pepsin and diastase. In 1867, Kühne suggested the name enzyme for the latter. Mechanists, such as Berthelot and Hoppe-Seyler, considered active agents that performed their catalytic function within the cells as ferments and when these agents were excreted outside the cell, they were called enzymes. Vitalists asserted that fermentations were necessarily intracellular and enzyme reactions were extra-cellular. The work of Buchner and Hahn on cell-free alcohol fermentation by yeast extracts dissipated this differentiation. The conclusion was that living protoplasm produced and carried enzymes, and they could be separated from the living tissue and retain catalytic activity, Waksman and Davidson *(1)*. This in the face of the belief by some researchers that various enzymes separated from the cell still retained a residue of vital force from the living material, Haldane *(5)*.

Vital forces aside, our present definition of enzymes still invokes the dictum that they are produced by living cells. A direct corollary of this concept was succinctly expressed by Tauber *(6)*, who noted that Ryshkov and Sukhov in 1928 had analyzed tobacco mosaic virus for the enzymatic activity of amylase, asparaginase, catalase, chlorophyllase, oxidase, peroxidase, phosphatase and protease. Their results were negative and Tauber concluded, therefore, that the virus is not a living thing. Enzymes are produced by living things; living things must produce enzymes.

One of the other major areas of speculation, agreement and dissent was concerned with the structure of an enzyme. Even a sketchy review of this subject indicates that the primary difficulty was related to the nature and relevance of proteins. They were ill understood, and this fact led to conceptual problems.

In 1922, Willstätter proposed his "Trager" or carrier theory. This viewpoint held that enzymes contain a smaller reactive group possessing a particular affinity for certain groupings in the substrate, thus accounting for the specificity of enzyme behavior. The reactive group, or enzyme proper, is considered to be attached to a colloidal carrier, and enzyme action is determined by the affinity of the active group for the substrate and by the colloidality of the entire aggregate. When the colloidal properties of the aggregate are destroyed, the activity of the enzyme disappears.

A particular colloidal carrier did not appear to be essential, but any suitable colloidal carrier could act as a protective colloid for the active group, Gortner and Gortner (7).

In a refinement of Willstätter's concept, Northrop (8) stated that it is possible that enzymes are similar to hemoglobin, containing an active group combined with an inert group. The active group may be too unstable to exist alone. Further, it is quite conceivable that a series of compounds may exist containing varying numbers of active groups combined with the protein, and that the activity of a compound depends on the number of active groups. It might be possible to attach more active groups to the inert group, thus increasing the activity above that of the original compound. This hypothetical complex does not differ much from that proposed by Willstätter and his co-workers, except that it supposes a definite chemical compound with a stabilizing moiety in place of an adsorption complex.

The interesting observation can be made that even Northrop's concept does not indicate a specific role for the protein in a catalytic sense. Despite this, a step had been taken in a positive direction since Kunitz and Northrop (9) in a related specu-lation indicated that if the native protein is merely a carrier for an active group, it is necessary to assume that the active group will become inactive when the protein is denatured and then will become active again when the protein reverts to the native condition. This is an early description of the reversible denaturation of an enzyme. It is also pertinent to point out that neither of these theories considered that parts of the protein (or colloid) itself might constitute the active site.

Quastel and Wooldridge (10) published an incisive set of thoughts on the relationship of adsorption, activation and specificity at the active center of an enzyme. They said that it was to be expected that a center possessing certain groups would adsorb a particular type of compound and another center with different groups would adsorb a different type of compound, i.e., the active center would evince a specificity of adsorption. But of the total number of molecules capable of being adsorbed at a particular center, only a few would be activated. The number of these would depend on the strength and nature of the polarizing field and the structure of the substrate molecules.

The mention of a "polarizing field" recalls that in 1919-1921, Barendrecht offered a radiation theory for enzyme activity: a molecule of urea absorbs radiation from urease and is hydrolyzed, Fearon (11).

We return to the currently accepted notion that (almost all) enzymes are proteins. Even in the 1930s and 1940s this fact rested uneasily in the minds of many scientists. J. B. S. Haldane (5) observed that preparations of gastric, pancreatic and hepatic esterases, and of yeast sacccharase and pancreatic amylase were obtained by

Willstätter and Bamann that were free from protein reactions (the biuret, Millon, ninhydrin and tryptophan reactions). Haldane concluded that the amount of protein in these preparations must have been small, and if, as many workers believed, the enzymes are all proteins, it was remarkable that the majority of the successful attempts to purify them led to the isolation of substances that were at least predominantly nonproteins, although the original material from which they were derived consisted largely of protein. Haldane also remarked that the purest preparations, which give no protein reactions, were still nondialyzable and, when analyzed, contained C, H, O, and N. Even more drastic and dramatic was the claim of Rao, et al. *(13)* that renin, a proteolytic enzyme, was probably less complex a structure than supposed, because it contained no detectable nitrogen, sulfur, or phosphorus. Their purest enzyme preparation contained carbon, hydrogen and oxygen along with some metals. Berridge *(13)* proposed that further work was required to clarify the contradictions inherent in the work of Rao, et al. and other investigators who stated the renin contained nitrogen.

Theorell *(14)* added more to the story by indicating that in 1926, Willstätter described his work on peroxidase purification, and it came to the point where ordinary analyses for protein, sugar, or iron indicated nothing: but enzyme activity remained. Willstätter concluded that enzymes might not contain any of these and did not belong to a known chemical class. He was included to believe that enzyme activity derived from a natural force. Even nothing has an explanation! The problem, of course, was that analytical methods were not sensitive enough to deal with these situations.

Levene *(4)* poignantly commented on the state of knowledge relevant to protein structure. He noted that the history of proteins, known from earliest time, was most discouraging. The term "protein" was introduced by Mulder in 1860, yet how little was known about the details of the structure of even a single protein, although the number of them in nature was endless.

On the more optimistic side, Haldane *(5)* commented that our definition of enzymes, produced by living cells, would be out of date when an enzyme was prepared synthetically. He tempered this prophetic remark by adding that his book would be out of date long before this occurred.

EARLY HIGHSPOTS

In an initial survey of the archeology of biocatalysis, some obvious highlight event may be recognized and noted (Table 1.). Because this table cannot show the fun, turmoil and intellectuality of scientific discovery a few highlights will be examined in greater detail.

Table 1. Selected great moments in early enzymology (15)

Dates	Investigators	Discoveries
1752	Reamur	Chemical aspect to gastric digestion
1783	Spallanzani	The same
1820	Planche	Plant extracts cause guaiacum blueing
1830	Robiquet	Amygdalin hydrolysis by bitter almond extract
1831	Leuchs	Ptyalin activity
1833	Payen and Persoz	Diastase (amylase) acitivity
1833	Beaumont	Food solvent other than HCL
1835	Fauré	Siningrinase activity
1835	Berzelius	Defined catalysis
1836	Schwann	Pepsin activity
1837	Liebig and Wöhler	Emulsin activity
1846	Dubonfaut	Invertase activity
1856	Corvisart	Trypsin activity
1867	Kühne	Proposed term *enzymes* for unorganized ferments
1894	Fischer	Enzyme stereoselectivity
1894	Takamine	Diastase patent
1897	Buchner	Unorganized ferments convert glucose to ethanol
1898	Duclaux	—*ASE* to indicate enzymes
1911-13	Bourquelot, Bridel and Verdon	Glucoside synthesis or degradation by enzymes in >80% ethanol, acetone
1915	Röhm	Tryptic enzyme patent
1922	Willstätter	Träger theory
1926	Sumner	Urease crystals
1930	Northrop	Pepsin crystals
1932	Northrop and Kunitz	Trypsin crystals

Confession of a Peroxidase Addict

For this author, studies on peroxidases have constituted a deep, longstanding interest. As suggested in the introductory section, a sense of the historical background can give added dimensions to research routine.

There is a close historical connection between guaiacum and peroxidases. In 1820, Planche observed that an extract of guaiacum turned blue in the presence of horseradish or milk. In 1855, Schönbein reported that the bluing reaction required three components: an extract of some plant or animal tissues, guaiacum and air or "oxygenated water," Paul *(16)*.

In 1494, Europe, in the throes of a syphilis epidemic, was blessed with the appearance of guaiacum, obtained from chips from the Caribbean trees *G. sanctum* and *G. officinale*, which "cured" syphilis when added to water in a steam bath, Munger *(17)*. The "cure" retained its credibility for nearly two centuries.

You can see how running an assay for peroxidase with guiaicol (present in guaiacum) is elevated to a higher level of enjoyment by such an anecdote of the past.

More specifically, a personal focus on the halogenating capacity of peroxidases has been a favorite area of concern. These enzymes can oxidize chloride, bromide and iodide. A book has derived from this interest, Neidleman and Geigert *(18)*. Unfortunately, the authors were not aware of a wayward dogma expressed by Harvey *(19)* in a paper on bioluminescence wherein he begins a sentence by proclaiming: "Because NaCl could not possibly be oxidized by photogenin (=luciferase)–or any other substance–. . . ." Thus, a whole family of enzymes was not considered to be conceivable. This phrase would have had a prominent place in the book's introduction, but this information was not accessible to the authors.

Gastric Juice: A Hearty Brew

Having just illustrated the charm of detail available in enzyme archeology, one is compelled to reveal more instances with the same appealing quality.

Gastric digestion has attracted the attention of a number of talented experimentalists, Carlson *(20)*. Reaumur in 1752 and Spallanzani in 1783 introduced food and sponges, contained in perforated metal or wooden capsules, into the stomachs of fish, birds and man. These capsules were subsequently recovered by means of attached strings, vomiting or rectal passage. It was demonstrated that gastric digestion involved chemical solution rather than physical rupture. In 1822, Alexis St. Martin endured an accidental shotgun blast that resulted in the development of a gastric fistula, Young *(21)*. In 1833, Beaumont did experiments on St. Martin and determined that gastric juice contained a food solvent other than hydrochloric acid. In 1839, Wassman showed this to be pepsin.

Arrhenius, the famous Swedish chemist and physicist, attempted to demonstrate that there was a mathematical relationship between the quantity of ingesta and the quantity of gastric juice produced. However, it remained for Pavlov's laboratory to report, in 1914, that the situation was more complex. It was concluded that, dependent on dietary intake, there were meat, bread and milk juices and these varied in their secretion period, volume, acidity and pepsin content. It is not clear, Carlson *(20)* that any of Pavlov's work was true, but it is comforting to realize that even 70 years ago gastric digestion gave rise to a bewildering and complex biochemical network.

ANTICIPATING AN ENZYME

Often, in the early days of enzymology, logic anticipated the existence of a biocatalyst before its existence was, in fact, confirmed. Two illustrations, out of many, will be considered: carbonic anhydrase and lipoxygenase.

The evolution of carbon dioxide from bicarbonate in the presence of blood solutions was an area of study in the 1920s and early 1930s. It had been tentatively concluded that the hemoglobin was the catalytic agent. However, Van Slyke and Hawkins *(22)* concluded that this was not the case and that some other catalyst was responsible. The problem centered about the fact that hemoglobin preparations, which showed carbon dioxide evolution, were contaminated with another substance that was subsequently identified and named carbonic anhydrase, Meldrum and Roughton *(23)*. To insert some sense of humanity into these academic flurries, Davenport *(24)*, a later worker with carbonic anhydrase, made the following pithy analysis of Roughton: "Roughton was not, to put it gently, an accomplished technician in the laboratory. My guess is that after Meldrum's death, when Roughton had to finish the work himself, he missed finding carbonic anhydrase in other tissues by sheer ineptitude."

The second example, that of lipoxygenase, is of a different sort because it graphically illustrates the marvelous but unsuspected anticipation of an enzyme and its active site by studies in chemical catalysis. Warburg *(25)* was studying the effect of iron in oxidation reactions of living cells. A segment of his research was devoted to a detailed investigation of model systems, involving iron, for the specific oxidation of amino acids, fructose and fatty acids. With regard to the fatty acids, Warburg noted that iron, combined with the sulfhydryl (—SH) group of cysteine, did not oxidize amino acids or sugar, but did specifically oxidize unsaturated fatty acids such as linoleic acid.

One train of thought that might be developed from such a finding is that these two entities, iron and —SH groups, might constitute important functionalities at the active site of an enzyme devoted to the oxidation of unsaturated fatty acids. Such an enzyme family is, in fact, that of the lipoxygenases. It has been demonstrated that iron and —SH groups are involved in their oxidative activity, Grossman, et al. *(26)*; Feiters, Veldink and Vliegenthart *(27)*. This is a case, then, in which the model system anticipated the enzyme, in contrast to many modern examples where the enzyme anticipates the model system or synzyme (artificial enzyme).

In selecting early heroes of modern industrial biocatalysis – a dangerous sport – significant names are bound not to be mentioned – but no one would argue with a list that began with Takamine, Wallerstein, Röhm, Boidin and Effront. This becomes evident when early patents are considered

Table 2 *(28)* presents selected early patents (1894-1938) related to the use of enzymes for commercial purposes. Scrutiny of the list begins to establish the idea that a creditable amount of interest existed in the period under discussion concerning the application of biocatalytic methods to commercial processes. There was a major

Table 2. Selected Early Enzyme Patents (28)

Inventors	Patent Number	Year	Enzyme	Title
J. Takamine	US 525,823	1894	Amylases	Process of making diastatic enzyme
J. Takamine	US 826,699	1906	Amylases	Diastatic substance and method of making same
O. Röhm	US 886,411	1908	Trypsin, steapsin	Preparation of hides for the manufacture of leather
J. Takamine	US 991,560	1911	Amylases	Enzyme
J. Takamine	US 991,561	1911	Amylases	Amylolytic enzyme
L. Wallerstein	US 995,820	1911	Malt Protease	Beer and method of preparing same
L. Wallerstein	US 995,823	1911	Proteases	Preparation of use in brewing
L. Wallerstein	US 995,824	1911	Pepsin	Method of treating beer or ale
L. Wallerstein	US 995,825	1911	Papain	Method of treating beer or ale
L. Wallerstein	US 997,826	1911	Bromelin	Method of treating beer or ale
L. Wallerstein	US 997,873	1911	Yeast protease	Method of treating beer or ale
O. Röhm	GER 283,923	1915	Pancreatin	Process for cleaning laundry of all types
S. Frankel	US 1,129,387	1915	Amylases	Manufacture of diatase
I. Pollak	US 1,153,640	1915	Amylases	Diastase preparations and method of making same
I. Pollak	US 1,153,641	1915	Amylases	Malt extract and method of making same
A. Boidin and J. Effront	US 1,227,374	1917	Amylases	Process for treating amylaceous substances
A. Boidin and J. Effront	US 1,227,525	1917	Amylases	Process of manufacturing diastases and toxins by oxidizing ferments
V.G. Bloede	US 1,257,307	1918	Amylase	Process of manufacturing vegetable glue
V.G. Bloede	US 1,273,571	1918	Amylase	Process of manufacturing vegetable glue
H.S. Paine and J. Hamilton	US 1,437,816	1922	Invertase	Process for preparing fondant or chocolate soft cream centers
J. Takamine	US 1,460,736	1923	Amylases, protease, lipase	Enzymic substance and process of making same
A. Boidin and J. Effront	US 1,505,534	1924	Amylases	Treatment of textile fabrics or fibers
Wallerstein Co.	UK 355,306	1931	Amylases, protease, lipase	Improvements in process of depilating hides
M. Wallerstein	US 1,854,353-5	1932	Amylases or papain	Method of making chocolate syrups
R. Douglas	US 1,858,820	1932	Amylases	Process of preparing pectin
L. Wallerstein	US 1,919,675	1933	Invertase	Invertase preparation and method of making the same
L. Wallerstein	US 2,077,447-9	1937	Proteases	Process of chillproofing and stabilizing beers and ales
L. Wallerstein	US 2,097,481	1937	Proteases	Rubber
L.. Wallerstein	US 2,116,089	1938	proteases	Deproteinization of rubber latex

involvement of the brewing industry, but the tanning, laundry, textile, candy, food and sweetener industries were also active.

Takamine was clearly a dominant force in the genesis of industrial enzymology. One of his favored microorganisms for enzyme production was *Aspergillus oryzae*. In 1914, he had these adoring comments for the fungus *(29)*: "Curiously enough this tiny and important hustler scarcely attracted attention in the Occident, and this fact made me determine to work for its introduction to industrial use in the United States."

Despite his determination, Takamine's attempts to replace malt by his fungal diastase in the alcohol fermentation industry was not a success. Underkofler, et al. *(30)* stated: "Use of mold preparations to replace malt in the fermentation industry was suggested by Takamine, and large-scale tests at the plant of Hiram Walker and Sons, Inc., in Canada in 1913, proved entirely successful, yields of alcohol being better than with malt. However, a slight off-flavor or odor was produced in the alcohol, and since the flavor is of paramount importance in beverage alcohol, Takamine's preparation has not found favor in the alcohol industry. Now, however, with the increasing interest in power alcohol, it would seem that a procedure similar to Takamine's should hold much promise for production of industrial alcohol."

The reason for the development of an off-flavor may be related, at least in part, to the existence of other enzymatic activities in the diastase preparation. The three recognized diastatic enzymes were identified as α-amylase, β-amylase and amyloglucosidase *(31)*. However, Harada *(32)* reported that the diastase preparation obtained by Takamine's method was a mixture of the following enzymes: alcohol oxidase, amidase, amylase, catalase, dextranase, ereptase, esterase, glycerophosphatase, inulase, invertase, lactase, lipase, maltase, peptase, phenolase, rennet, sulfatase and trypsin. Clearly, unless there was an intestinal or pancreatic contamination, rennet and trypsin were actually fungal enzymes with activities similar to those of the mammalian enzymes. Such an enzymatic polyglot would be likely to give an off-flavor. One such commercial product was appropriately called polyzime *(33)*.

An area of enzyme application with an atmosphere peculiar to itself was the tanning industry, in which animal hides were treated with the objective or producing acceptable leather for commercial use. Part of the process involved bating the hides in excrement tubs. Two types of tubs were in vogue: one redolent with dog dung, the other with a similar pigeon product. the first variety was the most prevalent. One aspect of the preparation and application of the doggy tub was described by Effront *(34)*: "Dog dung, generally imported from Asia Minor, is put in barrels and is sprinkled with a certain quantity of water in such a way as to moisten it and bring about a fermentation. After about a month, a certain quantity of this paste is taken, diluted in water, and soon afterward the clear liquid is decanted. It is in this maceration that the hides coming from barking and fleshing are plunged."

Eventually, the patents of Röhm (US 886411) and Wallerstein (UK 355306), which employed pancreatic extracts and bacterial liquors, respectively, helped to replace this technology.

ENZYME SOURCES *(1)(3)(5)(6)*

Enzymes for industrial applications were obtained from plant, animal and microbial sources as illustrated in Table 3 *(28)*. Beckhorn *(35)* noted in 1960 that the volume of enzymes produced from plant and animal sources exceeded that from microorganisms, but, he predicted that, for both technical and economic reasons, the role of microbial enzymes would become increasingly important. While this observation has been confirmed, another intriguing point is relevant here. Would the aficionados of biocatalysis have believed that, within 20-30 years, even plant and animal enzymes might be produced by microorganisms altered by the technology of molecular biology and genetic engineering?

In the 1950s, there was a technology revolution as well. Plants enzyme sources were grown in the field; animals as enzyme sources made their contribution as by-products of the meat industry; but enzymes from microorganisms were now produced in submerged fermentations rather than surface culture, which had been prevalent until the institution of this process improvement *(36)*.

What then were the major commercial applications for enzymes up to the early 1950s?

Table 3. Major Sources of Pre-1960 Commercial Enzymes (28)

Source	Enzyme
Microbial	
Yeast: *Saccharomyces cerevisiae*	Invertase
Saccharomyces fragilis	Lactase
Bacterial: *Bacillus subtilis*	Amylases, proteases
Micrococcus lysodeikticus	Catalase
Fungal: *Aspergillus flavus*	Amylases
Aspergillus niger	Glucosidases
Aspergillus oryzae	Proteases
Aspergillus niger	Catalase
Aspergillus notatum	Glucose oxidase
Aspergillus niger	Cellulase, lipase, pectinase
Plant	
Barley malt (*Hordum vulgare*)	Amylases (diatase)
Papaya (*Carica papaya*)	Protease (papain)
Pineapple (*Ananas commosus*)	Protease (bromelin)
Animal	
Beef pancreas (*Bos* sp.)	Trypsin, erepsin, amylase
Cattle and pig gastric extracts (*Bos* sp. and *Sus scrofa*)	Pepsin, lipase
Calf stomach (*Bos* sp.)	

ENZYME APPLICATIONS *(1)(6)*(34-42)

The use of enzymes in industrial processes of the early 1950s was already broad in scope, as indicated in Table 4 *(28)*. In fact, in scanning the information in the table, one might conclude that little has changed in 30-40 years. One obvious and major addition has been glucose (xylose) isomerase and its role in the production of high-fructose corn syrup. Much of the research in the past 30-40 years has concentrated on process improvement either by discovering and utilizing improved enzymes or by modifying the conditions of the enzymatic reaction and the subsequent isolation, purification, or presentation of products.

The introduction of enzymes into commercial processes has and often still does run the gauntlet of conservatism and tradition that is in the nature of many of the industries involved. Effront *(34)*, over 70 years ago, considered this issue in the cheese and beer domains. On cheese, he commented: "The manufacture of cheeses involves a certain number of practices imposed by ancient custom, which we formerly did not dare to dispense with for fear of coming to grief."

Apparently, the cheese industry did incorporate new technology and, therefore, escaped the incinerating ire of Effront. Not so the beer brewing industry.

Effront's concept of beer was as follows: "Beer must be considered primarily as a liquid food. In addition to taste and appearance, unquestionably very important factors, it must not only answer an immediate need of drinking, but must also bring to the organism the maximum nutrition. We can even say that the true excuse for its existence lies in the nourishment offered. On the other hand, beer should not be an expensive food, but a cheap one."

Effront's opinion of beer-makers was less kind: "When we study brewing a little more closely, we find an exaggerated respect on the part of the specialists in this branch for all the so-called natural products and practices which must be used in the course of this industry. The malt, the hops and the yeast are the three fundamental constituents that characterize beer. Without these, no beer is possible. The defenders of this opinion forget that beer is not a natural beverage, like wine, but that it forms, on the contrary, an artificial product created by man, and that nothing requires us to keep to the same formula."

"The brewing academies of Germany, Austria, America, etc., bear the stamp of institutions created for the maintenance of the brewing traditions. Traditions whose exclusive basis is the holy trinity, Hops, Malt, Yeast. The conservative spirit of the present science of brewing condemns it to sterility. If, in fact, we assume that only the three elements cited above can come into play in the manufacture of beer, the door of all investigation is forever closed."

One of the early problems with beer was that it formed a proteinaceous haze upon storage. Wallerstein *(43)* described the situation thusly: "At the turn of the century, almost all beer in this country was consumed as draught beer and little attention was paid to its stability and chillproof qualities, since it was intended to be consumed within a short period of time after it had left the brewery. with the developments in bottling, however, beer began to be shipped in bottles instead of

Table 4. Industrial applications of Enzymes (28)

Industry	Enzyme	Source*	Application
Baking	Amylase	M, P	Bread and cracker baking
	Protease	M	Bread and cracker baking
Beer	Amylase	M, P	Mashing
	Protease	A, M, P	Chillproofing
	Glucose oxidase	M	Oxygen removal
Carbonated drinks	Glucose oxidase	M	Oxygen removal
Cereals	Amylase	M, P	Precooked baby food
	Amylase	M, P	Breakfast food
Chocolate	Amylase	M	Syrup
Coffee	Pectinase	M	Bean fermentation
	Pectinase, hemicellulase	M	Concentrate
Condiments	Protease	A, M, P	Flavor ingredients
Confectionery	Invertase	M	Soft center, fondants
Dairy	Rennin	A	Cheese production
	Lipase	A	Cheese flavor
	Catalase	A, M	Milk sterilization with hydrogen peroxide
	Protease	A	Milk: off-flavor prevention
	Protease	A, M, P	Milk: protein hydrolysates
	Lactase	M	Whey concentrates, ice cream, frozen desserts, milk concentrates
	Glucose oxidase	M	Dried milk: oxygen removal
Distilled beverages	Amylase	M, P	Mashing
Egg	Glucose oxidase	M	Glucose removal
Fruit juice	Pectinase	M	Clarification, filtration, concentration
Laundry, dry cleaning	Protease, lipase, amylase	A, M	Spot removal
Leather	Protease	A, M	Bating
Meat	Protease	M, P	Meat tenderizing, casing tenderizing, condensed fish solubles
Molasses	Invertase	M	High-test molasses
Paper	Amylase	M, P	Starch modification for paper coating
Pharmaceutical	Amylase	M	Digestive aids
	Protease	A, M, P	
	Lipase	A	
	Cellulase	M	
	Protease (pepsin)	A	Nutritive peptones
	Proteases	A, M, P	Wound debridement
	Proteases	A, M, P	Injection for buises, inflammation
	Urease	P	Clinical test for urea
	Glucose oxidase plus peroxidase	M, P	Clinical test for diabetes
Photographic	Protease	M	Silver recovery
Starch	Amylase, amyloglucosidase	M	Corn syrup
	Amylase	M	Cold swelling laundry starch
Textile	Amylase	A, M, P	Fabric desizing
	Protease	A, M	Fabric desizing
Wallpaper	Amylase	M	Wallpaper removal
Wine	Pectinase	M	Pressing, clarification, filtration

* A = animal, M = microbiological, P = plant.

kegs over longer distances. As a result, the beer had to keep in good condition for considerably longer periods. While pasteurization, to which American brewers soon resorted, prevented biological breakdown of their products, the pasteurized, bottled beer, after a short time, however, developed a haze which greatly lessened its appeal to the eye. This problem became especially acute in view of the American custom of keeping beer for prolonged periods in the icebox and of serving it extremely cold."

So urgent was the demand throughout the brewing industry in the United States for a practical solution to this problem that in 1909 and again in 1910 the then U.S. Brewmasters Association offered two cash awards in a contest for the best papers on the causes of instability in finished, bottled beer. The submitted papers contained much valuable matter, with, however, a number of contradictory views. In any case, no satisfactory method of achieving a chillproof and stable beer was presented in any of these papers.

Subsequently, as indicated in Table 4, Wallerstein showed that the use of various proteases eliminated the haze problem. Effront *(34)*, whom we have experienced in his critical mode, was most positive about this advance made by Wallerstein: "In the course of the year 1913, the writer received from a brewery of New York two lots of beer of the same brew, one having undergone "wallerization," the other not having been treated. The analysis of these two lots, which were allowed to stand at the temperature of the laboratory, was made four months after the date of shipping. All the bottles treated had remained clear. On the other hand, the controls all contained a considerable deposit, and the liquid became turbid upon shaking and did not become clear again until after a prolonged standing. Further, the beers which had been treated were of a fresh, perfect taste; the others, on the contrary, had a flavor denoting a marked deterioration."

Early enzyme-based products were plagued by other characteristics that still skulk about to this day and, predictably, will continue to do so into the future: poor production control and unabashed adulteration.

Effront *(34)* notes the following with regard to the use of pepsin as a digestive aid: "In 1855, Corvisart first recommended the use of pepsin in therapeutics. Boudault, in 1857, prepared a powder with pepsin base which was much in fashion at that period. The success caused a great many limitations to appear, which soon led to a complete discrediting of this remedy. From suits entered against manufactures of pepsin in France at this time, it appears that there had been then delivered to the public products completely inactive, unclean, and dangerous."

Another protease, papain, had similar irregularities. In 1884, Martin *(44)* reported the following philosophical and practical considerations: "*Papain* was the name given by Wurtz to the proteolytic ferment found in the juice of the unripe fruits, of the stem and leaves of *Carica papaya*, a plant indigenous to the East and West Indies. The occurrence of such a ferment in the vegetable kingdom is in itself remarkable, and may give rise to important considerations on the researches of Darwin and others on the assimilation of animal food by carnivorous plants. The leaves and juice have long been known in the native countries of the plant to have an action on uncooked meat, which is rendered tender and subsequently rotten."

Of further historical interest is that the chemist Vanquelin *(45)*, in 1802, described papaya juice as blood without pigment due to its considerable digestive activity.

From this colorless observation it was only a short step to produce a commercial meat tenderizer using papain. However, there were some tough issues to face, as discussed by Mansfield *(46)*: "People looking about for ways and means for reducing the high cost of living have overlooked papain. A round steak treated with a solution for an hour or so before cooking becomes as tender and palatable as the best tenderloin. Papain is too valuable a drug to be discredited and forced into disuse merely because of an insatiable desire for profits. During the past few years it has been a common practice to adulterate papain."

One of the common ways to adulterate papain was to incorporate it into unleavened bread or to completely eliminate the papain, selling only the bread. Mansfield *(16)* described the economic advantages of such skullduggery: "Rice bread costs about eight cents per pound. (No yeast or salt is used in its preparation.) The same eight-cent bread when sold as papain brings over two dollars per pound, thus netting a profit to the adulterator of about 2400 percent. Schemes for extracting gold from sea water or selling gold bricks are Christian acts compared to the getting-rich-quick, and the safe (to date) practice of selling rice bread for papain."

It is worth noting that papain had many other applications in the United States, including use in the treatment of peritoneal and other adhesions, chronic dyspepsia, gastritis, diphtheria and a recommendation for use in eczema *(47)*.

It was mentioned above that not all was well with pepsin as a digestive aid. In another example concerning a mixture of enzymes in solution used as a digestive aid, Sollman *(48)* made the following incisive points: "If the solution is acid (as in the Elixir Digestivum Compositum of the National Formulary and most of the proprietary digestant mixtures), the trypsin and diastase will be destroyed; if it is alkaline, the pepsin and diastase will disappear, and if, as a last resort, it is made neutral, the pepsin will destroy the diastase, and the pepsin, in its turn, will be digested by the trypsin. At room temperature the process will be somewhat slower than in the thermostate, but the final result, in a very short time, will be the same."

On the other hand, it is only fair to balance the case by mentioning an enzyme product including pepsin that was both remarkable in its technical aspects for the time and apparently and mercifully efficacious. In this regard, it was reported by Nemetz *(49)* that favorable clinical results had been obtained with a new enzyme product marketed by "Norgine A. G. Prague" under the trade name "Enzypan." This product contained an outer shell that dissolved in the stomach, releasing gastric enzymes, while the pancreatic enzymes, trypsin, lipase, and amylase, were incorporated in the inner shell of the tablet, which released its contents into the duodenum. This represents an impressive example of any early timed-release formulation.

A few remarks about teeth and urine will serve to close this section on enzyme applications. The enzyme urease converts urea to ammonia and carbon dioxide and is a means of producing an alkaline medium. This controlled release of ammonia was a basis for a toothpaste containing urease. Further, Smythe *(39)* pointed out that another toothpaste, as well as a cleaning solution for false teeth, contained a protease and, in principle, these are analogous to the desizing applications noted in Table 4 for the textile industry. These preparations were, then, desizers for incisors.

Smythe in discussing urease, made a suggestion that had technical merit (not really) and no aesthetic appeal (really): "There are other potential applications for this enzyme such as the interesting one of recovering water from urine under conditions of severe water deprivation, as might be imposed by desert warfare."

Even without a major market position for this last potential application, it is clear that the industrial use of enzymes in the first half of the 20th century was varied and intimately associated with many profitable processes and products.

WATER-INSOLUBLE ENZYMES CAN WORK TOO

The interactions of enzymes and water-insoluble matrices in enzyme immobilization is an area of active research at present. The past also had its moments. In 1910, Starkenstein reported on the adsorption of amylase on its substrate starch, Hais (50). He demonstrated that dialyzed amylase could be adsorbed on water-insoluble starch. The complex could be recovered, washed, and then activated to cause starch degradation in the presence of chloride ion. He ascribed the formation of the amylase-starch complex to purely physical forces. The enzyme activity did not depend on the presence of a soluble enzyme.

In a continuation of this theme of enzyme immobilization by adsorption, Nelson and Griffin (51) demonstrated that yeast invertase could be absorbed on charcoal and retain its catalytic activity. Levene and Weber (52) showed that nuccleosidase adsorbed on kaolin retained its activity and could not be extracted from the matrix by a variety of aqueous solutions.

CHEMICAL MODIFICATION OF ENZYMES

Early workers were not only aware that enzyme activity did not necessarily depend on water solubility but that, in addition, enzyme structure could be modified, without total elimination of catalytic activity. These were studies designed to investigate the importance of particular functionalities in determining enzyme activity. Herriott and Northrop (53) and Tracy and Ross (54) worked with derivatives of pepsin using ketene and carbon suboxide to acetylate and malonylate, respectively, the enzyme. Both amino and phenolic hydroxyl groups could be altered. It was determined that, whereas loss of amino groups did not appear to affect enzyme activity, loss of tyrosyl hydroxyl groups inactivated the enzyme. It was further claimed that malonylation of the lysyl amino moieties did not alter pepsin specificity: that amino groups could be replaced by carboxyl groups without effect.

As a final illustration of studies related to modification of enzyme structure, Gjessing and Sumner (55) showed that the "natural" iron porphyrin present in horseradish peroxidase could be replaced by a manganese porphyrin and the resultant, regenerated catalyst had 20-30% of the activity obtained when iron porphyrin was used in its place.

The work noted in this and the preceding section illustrates early enzymologists realized that enzymes, as isolated from nature, could be chemically and physically altered without a major loss in activity.

ENZYMES IN ORGANIC SOLVENTS CAN GO EITHER WAY

Research related to the catalytic activity of enzymes in the presence of high concentrations of organic solvents has reached explosive intensity in the past decade. In reporting on this, the statement has been made that, in the early 1980s, Klibanov found that lipase activities were retained in nonaqueous solutions, Gillis *(56)*. The inference that this constituted a pristine discovery is inaccurate by a margin of over 50 years, and work with another enzyme in organic solvents was carried out over 75 years ago.

In a extraordinary series of papers, Bourquelot and co-workers (Bourquelot and Bridel (57-61); Bourquelot and Verdon (62-63)) studied the action of emulsin (containing β-glucosidase) in organic solvents. It was reported that this enzyme from almonds, when placed in suspension in 85% alcohol containing glucose in solution, caused at ordinary temperature the combination of sugar and alcohol with formation of β-ethylglucoside. Furthermore, it was found that emulsin as a powder had a hydrolytic activity on dissolved glucosides in neutral liquids such as acetone and ethyl ether, although these liquids did not dissolve a trace of enzyme.

These studies were early demonstrations that enzymes were involved in equilibrium reactions and could go either way depending on conditions. Further, it established that enzyme activity in organic solvents was feasible.

With regard to investigations on lipase activity in the presence of organic solvents, the following may be noted:

Sym *(64)* showed that in the case of pig pancreatic lipase, the application of organic solvents, practically insoluble in water, gave much higher yields of esters (often >95%) than the use of aqueous systems, owing to the fact that the products of reaction were removed from the medium where the reaction proceeds. In addition, the nonaqueous phase may be regarded as a reservoir of the components of the reaction, supplying substrates to the aqueous phase. The action of the enzymic preparations was studied, in most cases, in systems containing 1 M n-butanol alcohol, 0.43 M butyric acid and benzene. In a number of studies, the enzyme preparation was an acetone powder containing 8-12% water. Sperry and Brand *(65)* showed cholesterol esterification using a similar enzyme preparation in ≤90% carbon tetrachloride. With butyric acid, 65.2% esterification occurred; with palmitic acid, 30%; and with oleic acid, 40%.

It is evident that the explosive expansion of research on enzyme reactions in organic solvents began building energy over 75 years ago. The literature awaits further excavation.

ENZYME SPECIFICITY: AN INTELLECTUAL CALDRON

In a landmark paper on the effect of configuration on enzyme activity, Fischer *(66)* expressed a number of opinions that ring true after almost 100 years:

". . . bezweisle ich ebensowenig wie die Brauchbarkeit der Enzyme für die Ermittlung der Configuration asymmetrischer Substanzen."

[I have little doubt about the usefulness of enzymes for the determination of the configuration of asymmetric substances.]

"Noch wichtiger für dieselbe aber scheint mir der Nachweis zu sein, dass der früher vielfach angenommene Unterschied zwischen der chemischen Thätigketi der lebenden Zelle und der Wirkung der chemischen Agenten in Bezug auf moleculare Asymmetrie thatsächlich nicht besteht."

[Also important it appears to me is that the evidence shows that the earlier repeated and accepted difference between the chemical activity of the living cell and that of chemicals with regard to asymmetric molecules does not in fact exist.]

"Aber shon genügen die Beobachtungen, um principiell zu beweisen, dass die Enzyme bezüglich der Configuration ihrer Angriffsobjecte ebenso wählerisch sind, . . ."

[But the results already suffice to prove the principle that enzymes are fussy about the configuration of their object of attack.]

"Um ein Bild ze gebrauchen, will ich sagen, dass Enzym and Glucosid wie Schloss und Schlüssel zu einander passen müssen, um eine chemische Wirkung auf einander ausüben zu können."

[To use an image, I would say that the enzyme and glucoside must fit each other like a lock and key to be able to carry on a chemical reaction on each other.]

The future confirmed the genius of Fischer.

It is to be emphasized that the lock-and-key metaphor is invaluable as an image for those learning to think of enzyme-substrate interactions, but it should also be emphasized that modern work has built on this rigid portrait and added dimensions of conformational flutter. Enzyme and substrates stand still for no reaction.

Thunberg *(67)* presented another indication that enzymes have specificity by observing that individual enzymes in a mixture could be differentiated by their different temperature responses, especially their thermostability. Therefore, he reasoned there are a variety of distinct enzymes catalyzing a variety of reactions. In 1921, Dakin commented that Thunberg's argument was rather unconvincing, but Bernheim *(68)* offered an analogous suggestion to that of Thunberg's by saying that one type of evidence for enzyme specificity was in the separation of the enzyme from the tissue:

"This evidence is fairly conclusive for the separation depends on differences in physical properties of the enzymes, i.e., the centers responsible for the activations are attached to different colloids. It would be difficult to explain these facts on the assumption of one enzyme in the tissue which was originally capable of effecting all the activations but which during the process of extraction has been so altered as to appear specific for one substrate. Four dehydrases have been separated in this way: the succinic, lactic, citric and xanthine."

Thunberg argued for specific enzymes on the basis of thermostability and Bernheim because of separation individuality.

It is more and more common to read in the present literature of "unexpected" catalytic activity of known enzymes, Klibanov *(69)*. From this viewpoint, it is enlightening to consider how the "expected" catalytic activity of two specific enzymes metamorphosed over a period of years. The two enzymes are xanthine oxidase and urease.

Morgan, Stewart and Hopkins *(70)* stated that the enzyme xanthine oxidase showed marked specificity, accepting xanthine and hyperoxanthine as substrates but not guanine, caffeine, uracil, thymine, or cytosine, for example. The authors were, however, skeptical about reports that the enzyme could also oxidize aldehydes such as acetaldehyde. They thought it unlikely that the enzyme would extend its activity to such compounds. Dixon *(71)* was equally unsure that xanthine oxidase could oxidize both purines and aldehydes:

"With regard to the case of aldehyde, it cannot be maintained that it is definitely established that the oxidations of aldehyde and purines are due to the same enzyme. If they are not, then *both* enzymes must be remarkably specific. On the other hand, there appears to be a fairly strong balance of evidence in favor of identity, although a certain part of this can be explained away by supposing that we have two different enzymes adsorbed on the same colloid."

This is, indeed, a classic case of inconclusiveness.

Keilen and Hartree *(72)* concluded, on the basis of the work of others, that the oxidation of purines and aldehydes was definitely due to a single enzyme, xanthine oxidase. Coming nearly full circle after nearly 50 years, Dixon and Webb *(73)* stated that xanthine oxidase had a rare, dual specificity, oxidizing purines and aldehydes. It is delightful to report that these authors actually wrote *duel* specificity, as if to indicate subconsciously that arguments over enzyme specificity can be barbed on occasion, as in this case wherein two camps existed: the single enzyme and two enzyme forces.

The background related to the specificity of urease is simpler in the sense that only urea and closely related analogues were studied rather than diverse substances such as purines and aldehydes. In the case of xanthine oxidase, Armstrong and Horton (74) concluded, quite correctly, based on the range of substrates examined, that urease was specific for urea. Other substrates examined were biuret and various methyl and ethyl derivatives of urea. The authors also suggested that the enzyme must "correspond very closely in structure with the hydrolyte urea with which alone it is in correlation."

Sumner (75) succeeded in preparing the first crystalline enzyme, urease. One can sense the stamp of priority in his opening paragraph:

"After work both by myself and in collaboration with Dr. V. A. Graham and Dr. C. V. Noback that extends over a period of a little less than 9 years, I discovered on the 29th of April a means of obtaining from the jack bean a new protein which crystallizes beautifully and whose solutions possess to an extraordinary degree the ability to decompose urea into ammonium carbonate."

Thus, it is established that Sumner had a vested interest in urease, and so it is of more than passing interest to quote the following from Sumner and Somers (76): "Urease is an enzyme that is absolutely specific. It acts only upon urea and nothing else." Therefore, it might be assumed that his is an example of a known, purified enzyme with no hope for unexpected substrates. There is always hope! Gazzola, Blakely and Zerner (77) reported other compounds hydrolyzed by urease were N-hydroxyurea, semicarbazide (N-aminourea) and N,N'dihydroxyurea, while carbazide (N,N'diaminourea) N-methylurea, N-butylurea, N-methoxyurea, N-(2-aminoethyl)urea, 2-imidazolidinone, biuret, 1-phenylsemicarbazide, ethyl carbamate and carbamoyl azide were among many derivatives that were not hydrolyzed.

Sumner, as the first crystallizer of an enzyme, deserves a biographical sketch. Laskowski (78) offered some first-hand insights into the person. He was a millionaire descendent of a Mayflower family. He lost an arm at 15 and won a Nobel prize at 59. His lectures were considered dull and bloated with detail. His mind was a living encyclopedia of biochemistry.

The above brief and superficial discussion of enzyme specificity suggests that unexpected substrates may be unexpected but not a surprise because we still know relatively little about the total potential of any given enzyme under a variety of reaction conditions, given an exception now and again.

DEVELOPMENTS OVER TIME

One of the obvious advances made in enzymology during the past 50 years is the development of high technology analytical capabilities. When the pioneers of enzymology were building their proteinaceous pyramids, they had only a vague idea of what the building material was. Today, with our molecular microscopes, we nearly know it all. One brief illustrative example is in the conversion of chymotrypsinogen to chromotrypsin.

Kunitz and Northrop (9) reported that:

"The transformation of chymo-trypsinogen into chymo-trypsin is accompanied by a change in optical activity and a slight increase in amino nitrogen. There is no detectable non-protein nitrogen fraction formed nor is there any significant change in molecular weight. The reaction, therefore, is probably an internal rearrangement, possible due to the splitting of a ring."

In a subsequent publication, Herriott and Northrop (53) refined this interpretation by suggesting that the rupture of a peptide bond was involved. Given the state of their art, this was an incisive conclusion. However, the transformation was later summarized in more graphic detail by Stryer (79):

"Some conformational changes that have been elucidated include:
1. Cleavage of peptide bond at $Arg^{15}Ile^{16}$.
2. $-NH_2$ of Ile^{16} interacts with $-COOH$ of Asp^{194}.
3. Other changes follow:
   A. Met^{192} moves from interior to surface.
   B. Residues 187 and 193 become extended.
   C. Substrate specificity pocket is thus created.

The contrast is shocking when it is realized that the best the early workers could do was to implicate the cleavage of an unidentified peptide bond, whereas later workers could deal with the transformation of specific molecular domains. The intellect was willing but information and technique were wanting.

PUBLICATION VELOCITY: A SURVIVAL FACTOR

The insatiable drive to establish scientific priority survives unabated in these times. It is, however, not a product of our times. Two examples from the past will illustrate this with blinding clarity. The first might be entitled: "At 4:01 p.m. on August 22, 1888, I knew I had a ferment." Green (80) reported an experiment designed to demonstrate that a ferment existed in extracts of seeds of *Ricinus communis* (Castor oil plant) that could produce fatty acids from castor oil:

"Tube F was prepared by mixing the extract and the emulsion of castor oil in the proportions given above, and was put into an incubator at 12:30 o'clock on August 22, 1888. A boiled control was put with it, labelled F_1 neutral."

The experiment was successful and priority was ensured.

The second example is by Dakin *(81)* who had submitted a paper on the conversion of acetoacetic acid to β-oxybutyric acid in cats and dogs. He comments:

"This fact was recorded in a short paper sent on March 16 to the *Journal of the American Medical Association*, which, however, not withstanding a promise of prompt publication, did not appear until April 29. In the meantime, a paper appeared by Blum (March 29) in which the same fact was established by slightly different methods."

The Blum paper was published in a different journal and even more irritating was the fact, noted by Dakin, that Blum made reference to similar experiments by someone named Maase, the publication of which was not yet accessible to Dakin.

AN EXAMPLE OF SERENDIPITY IN SCIENTIFIC DISCOVERY

Fleming had a prepared mind in 1929, when he reported the discovery of penicillin. Earlier in his career, he had another incident of the prepared mind: his discovery of lysozyme. Fleming *(82)* described the event:

"In this communication I wish to draw attention to a substance present in the tissues and secretions of the body, which is capable to rapidly dissolving certain bacteria. As this substance has properties akin to those of ferments I have called it a 'Lysozyme,' . . .

In the first experiment nasal mucus from the patient, with coryza, was shaken up with five times its volume of normal salt solution, and the mixture was centrifuged. A drop of the clear supernatant fluid was placed on an agar plate, which had previously been thickly planted with *M. lysodeikticus*, and the plate was incubated at 37°C for 24 hours, when it showed a copious growth of the coccus, except in the region where the nasal mucus had been placed. Here there was complete inhibition of growth, and this inhibition extended for a distance of about 1 cm. beyond the limits of the mucus."

What makes this occurrence even more intriguing is the claim that the patient was Fleming, Stryer *(79)*:

"In 1922, Alexander Fleming, a bacteriologist in London, had a cold. He was not one to waste a moment, and consequently used his cold as an opportunity to do an experiment. He allowed a few drops of his nasal mucus to fall on a culture plate containing bacteria. He was excited to find some time later that the bacteria near the mucus had been dissolved away and thought that the mucus might contain the universal antibiotic he was seeking."

TIME AND THE CYTOCHROMES: TERMINATION AND RESUSCITATION

Destructive criticism wanders through the pages of enzymology. A particularly devastating instance involves the cytochromes. MacMunn *(83)* described what he believed to be novel respiratory pigments in muscle and other tissues of animals. He used the terms "myohaematin" and "histohaematin" to identify them. Levy *(84)* repeated the work of MacMunn and regarded the substances of MacMunn as haemochromogen, derived from hemoglobin. An exchange of disagreements was initiated by MacMunn *(85)* and Hoppe-Seyler *(86)* who sided with Levy. MacMunn *(87)*, made one last attempt to salvage his work. His effort was discredited when Hoppe-Seyler *(86)* appended a terse editorial note at the end of this publication saying that MacMunn argued with no new facts and further discussion was not to occur: there was no reason for pigeon muscle to contain a special pigment.

MacMunn's new respiratory pigment was forgotten. Fortunately, Keilen *(87)* revived the subject. He renamed MacMunn's pigment "cytochrome" and demonstrated that cytochrome was distinct from muscle hemoglobin; that both pigments could be readily detected in the same muscle of a bird or mammal; and that cytochrome also occurred in yeast, bacteria and higher plants.

EFFECT OF TEMPERATURE ON FAT UNSATURATION

Neidleman *(88)* published a broad review of the effects of temperature on lipid unsaturation. An oversimplified, single sentence summary would be as follows:

The colder the living system, the more unsaturated are its lipids; the warmer the living system, the less unsaturated are its lipids and enzymatic unsaturation is intimately involved in these effects.

In preparing the review, a wonderful experiment described by Henriques and Hanson *(89)* was not rediscovered. These investigators acquired three young pigs from the same litter. Each was given special and individualized attention. One pig was kept in a room at 30-35°C, while the others were kept in pens at about 0°C (the experiment was completed during the winter). Of the latter two pigs, one animal had a sheepskin, the wood side turned inward, placed on its back, stomach and sides. All three animals were fed only a corn diet. After 2 months all three animals were slaughtered and the skin fat, as well as the kidney and omental fat were examined.

The results indicated that the 20-35°C pig had the least unsaturated fat, the unadorned pig at 0°C had the most unsaturated fat, and the pig in a blanket was intermediate in lipid unsaturation. Enzymes involved in determining lipid unsaturation are affected by temperature.

CONCLUSIONS

In 1926, Waksman and Davison *(1)* wrote: "Life is just one enzyme reaction after another." On the other hand, it is equally true that industrial catalysis is not just one enzyme reaction after another. Many commercial reactions are chemically catalyzed, not enzymatically catalyzed. However, this historical chapter and the state-of-the-art chapters to follow illustrate that industrial biocatalysis has been around for a long time and will become more entrenched in the future as it gains further respect as an economically acceptable solution to many industrial problems.

This chapter might have been expanded by including examples wherein the biocatalytic agent is a live, intact cell rather than a quasi- or completely purified enzyme. The discussion would, then, have included such items of commercial interest as citric acid (90-91), gluconic acid *(92)*, acetone and butanol *(93)* and vinegar *(94)*. But it did not.

It seems appropriate that this historical chapter should give the final words to those reviewing biocatalysis in the period 1939-1960 rather than to a contemporary voice. What were their thoughts about the future?

1. "As our knowledge regarding enzymes advances, it is not too much to anticipate that in time more and more of the enzymes are likely to be put to work industrially." *(37)*
2. "Through the ceaseless efforts of scientists and technologists, enzymes have been successfully harnessed to the needs of industry. Progress has been furthered, not only by the knowledge of the behavior of enzymes, which research has expanded, but by the development of enzyme-containing preparations designed to meet the requirements of specific processes. The increasing complexity of modern industry, it is believed, will bring about a corresponding increase in new and interesting uses for this fascinating group of organic catalysts." *(38)*
3. "Many industrial problems have been solved by the proper application of enzyme chemistry; others await improvements or solution. Indeed the fields for exploration offered by catalytic microorganisms and practical enzymology are almost unlimited." *(6)*
4. "Finally, it must be pointed out that the synthesis of our principal foods – sugars (starches) and amino acids (proteins) – is carried out enzymatically by plants and animals. The simple building stones for many of these processes are sometimes available, and a few of the enzymes involved in these syntheses have been studied. It is not impossible to visualize an era in which the enzymatic synthesis of our common foods could be undertaken." *(95)*
5. "Currently much enzyme research is underway by various industries including enzyme manufacturers. Such research is devoted to finding new and improved methods for using enzymes, to improving yield of industrial microbial enzymes, and to finding new enzymes for industrial purposes.

Continually increasing usage of old and new enzymes will result form such research." *(1)*

6. "It is generally conceded that we have only begun to tap the full potential of enzymes. The extensive research now in progress will bear fruit, making new systems and new purified forms available. Cooperative research with industrial users of enzymes should continuously uncover new areas in which enzymes can serve a useful role." *(35)*

7. "The recent growth in the industrial and medical applications of enzymes has resulted from greatly improved methods for their production, especially from plant and microbial sources. In no small degree it has also evolved as a result of our deeper understanding of the chemical and physical properties of enzymes and of the kinetics of the total enzyme-substrate system, with special reference to the role of activators and inhibitors. This fundamental knowledge of enzyme systems will inevitably lead to major growth in the industrial and medical applications of enzymes in the near future." *(41)*

These statements have an unmistakable similarity to contemporary pronouncements.

LITERATURE CITED

1. Waksman, S. A.; Davison, W. C. *Enzymes, Properties, Distribution, Methods and Applications*; Williams & Wilkins: Baltimore, MD, 1926.

2. Fruton, J. S.; Enzyme. In *The Encyclopedia Americana*; Grolier: Danbury, CT, 1981; *10*, 489-493.

3. Fruton, J. S.; Simmonds, S. *General Biochemistry*, 2nd ed.; John Wiley & Sons: New York, 1958.

4. Levene, P. A. *Science* 1931, *74*, 23-27.

5. Haldane, J. B. S. *Enzymes*; M.I.T. Press: Cambridge, MA, 1930; Longmans, Green and Co.: New York, Reprint 1965.

6. Tauber, H. *Enzyme Technology*; John Wiley & Sons: New York, 1943.

7. Gortner, R. A., Jr.; Gortner, W. A. *Outlines of Biochemistry*, 3rd ed.; John Wiley & Sons: New York, 1949.

8. Northrop, J. *J. Gen. Physiol.* 1930, *13*, 739-766.

9. Kunitz, M.; Northrop, J. H. *J. Gen. Physiol.* 1935, *18*, 433-458.

10. Quastel, J. H.; Wooldridge, W. R. *Biochem. J.* 1928, *22*, 689-702.

11. Gearon, W. R. *Biochem. J.* 1923, *17*, 84-93.

12. Rao, C. N. B.; Rao, M. V. L.; Ramaswamy, M. S.; Subrahmanyan, V. *Curr. Sci.* 1941, *4*, 179-186.

13. Berridge, N. J. *Biochem. J.* 1945, *39*, 179-186.

14. *Reflections on Biochemistry in Honor of Severo Ochoa*; Kornberg, A.; Horecker, B. L.; Cornudella, L.; Oro, J., Eds.; Pergamon Press: New York, 1976; pp. 57-63.

38

15. Neidleman, S. L. In *Biocatalysis*; Abramowitz, D. A., Ed.; Catalysis Series, special print; Van Nostrand Reinhold: New York, 1990; pp 1-24.
16. Paul, K. G. *J. Oral Pathol.* 1987, *16*, 409-411.
17. Munger, R. S. *J. History Med.* 1949, *4*, 196-229.
18. Neidleman, S. L.; Geigert, J. *Biohalogenation: Principles, Basic Roles and Applications*; Ellis Horwood Ltd.: Chichester, England, 1986.
19. Harvey, E. N. *J. Gen. Physiol.* 1918, *1*, 133-145.
20. Carlson, A. J. *J. Physiol. Rev.* 1923, *3*, 1-40.
21. Young, S. *New Scientist* 1985, *108*, 24-27.
22. Van Slyke, D. D.; Hawkins, J. A. *J. Biol. Chem.* 1930, *87*, 265-279.
23. Meldrum, N. U.; Roughton, F. J. W. *J. Physiol. (Lond.)* 1933, *80*, 113-142.
24. Davenport, H. W. *Ann. NY Acad. Sci.* 1984, *429*, 4-9.
25. Warburg, O. *Science* 1925, *61*, 575-582.
26. Grossman, S.; Klein, B. P.; Cohen, B; King, D..; Pinsky, A. *Biochim. Biophys. Acta* 1984, *793*, 455-462.
27. Feiters, M. C.; Veldink, G. A.; Vliegenthart, J. F. G. *Biochim. Biophys. Acta* 1986, *870*, 367-371.
28. Neidleman, S. L. In *Biocatalysis For Industry*; Dordick, J. S., Ed.; Topics in Applied Chemistry; Plenum Press: New York, 1991, pp 21-33.
29. Takamine, J. *J. Ind. Eng. Chem.* 1914, *6*, 824.
30. Underkofler, L. A.; Fulmer, E. I.; Schoene, L. *Ind. Eng. Chem.* 1939, *31*, 734.
31. Langlois, D. P. *Food Technol.* 1953, *7*, 303.
32. Harada, T. *Ind. Eng. Chem.* 1931, *23*, 1424.
33. Smyth, H. F.; Obold, W. L. *Industrial Microbiology: The Utilization of Bacteria, Yeasts and Molds in Industrial Processes*; Williams & Wilkins Co.: Baltimore, MD, 1930; p 256.
34. Effront, J. *Biochemical Catalysis in Life and Industry*, Trans. by S. C. Prescott; John Wiley and Sons, Inc.: New York, 1917.
35. Beckhorn, E. J. *Wallerstein Lab. Commun.* 1960, *23*, 201.
36. Underkofler, L. A.; Barton, R. R.; Fennert, S. S. *Appl. Microbiol.* 1958, *6*, 212.
37. Wallerstein, L. *Ind. Eng. Chem.* 1939, *31*, 1218.
38. Gale, R. A. *Wallerstein Lab. Commun.* 1941, *4*, 112.
39. Smythe, C. V. *Econ. Bot.* 1951, *5*, 126.
40. Beckhorn, E. J.; Labbel, M. D.; Underkofler, L. A. *J. Agr. Food Chem.* 1965, *13*, 30.
41. Sizer, I. W. *Adv. Appl. Microbiol.* 1964, *6*, 207.
42. Underkofler, L. A. *Bioeng. Food Process.* 1966, *62*, 11.
43. Wallerstein, L. *Wallerstein Lab. Commun.* 1956, *19*.
44. Martin, S. H. C. *J. Physiol. (London)* 1884, *5*, 213.
45. Vanquelin, L. N. *Ann. Chim. Phys.* 1802, *18*, 267.
46. Mansfield, W. *J. Am. Pharm. Assoc.* 1914, *3*, 169.
47. Kwang, K.; Ivy, A. C. *Ann. N. Y. Acad. Sci.* 1951, *54*, 161.
48. Sollman, T. *J. Am. Med. Assoc.* 1907, *48*, 415.
49. Nemetz, C. *Dtsch. Med. Wochenschr.* June 22, 1928, p. 1047.
50. Hais, I. M. *J. Chromatogr.* 1986, *373*, 265-269.

51. Nelson, J. M.; Griffin, E. G. *J. Chem. Soc.* 1916, *38*,1109-1115.
52. Levene, P. A.; Weber, I. *J. Biol. Chem.* 1924, *60*, 707-715.
53. Herriott, R. M.; Northrop, J. H. *J. Gen. Physiol.* 1934, *18*, 35-67.
54. Tracy, A. H.; Ross, W. F. *J. Biol. Chem.* 1942, *146*, 63-68.
55. Gjessing, E. C.; Sumner, J. B. Arch. B*iochem. Biophys.* 1942, *1*, 1-8.
56. Gillis, A. *J. Am. Oil Chem. Soc.* 1988, *65*, 846-850.
57. Bourquelot, E.; Bridel, M. *J. Pharm. Chim.* 1911, *4*, 385-390.
58. Bourquelot, E.; Bridel, M. *Compt. Rend. Acad. Sci.* 1912, *154*, 944-946.
59. Bourquelot, E.; Bridel, M. *Compt. Rend. Acad. Sci.* 1912, *154*, 1375-1378.
60. Bourquelot, E.; Bridel, M. *Compt. Rend. Acad. Sci.* 1912, *155*, 86-88.
61. Bourquelot, E.; Bridel, M. *Compt. Rend. Acad. Sci.* 1912, *155*, 437-439.
62. Bourquelot, E.; Verdon, E. *Compt. Rend. Acad. Sci.* 1913, *156*, 1264-1266.
63. Bourquelot, E.; Verdon, E. *Compt. Rend. Acad. Sci.* 1913, *156*, 1638-1640.
64. Sym, E. A. *Biochem. J.* 1936, *30*, 609-617.
65. Sperry, W. M.; Brand, F. C. *J. Biol. Chem.* 1941, *137*, 377-387.
66. Fischer, E. *Chem. Ber.* 1894, *27*, 2985-2993.
67. Thunberg, T. *Skand. Arch. Physiol.* 1920, *40*, 1-91.
68. Bernheim, F. *Biochem. J.* 1928, *22*, 1178-1192.
69. Klibanov, A. M. *Basic Life Sci.* 1983, *8*, 497-518.
70. Morgan, E. J.; Stewart, C. P.; Hopkins, F. G. *Proc. Roy. Acad. Sci. Lond. Ser. B. Biol.* 1922, *94*, 109-131.
71. Dixon, M. *Biochem. J.* 1926, *20*, 703-718.
72. Keilen, D.; Hartree, E. F. *Proc. Roy. Acad. Sci. Lond. Ser. B. Biol.* 1935, *119*, 114-140.
73. Dixon, M.; Webb, E. C. *Enzymes*; Academic Press: New York, 1979.
74. Armstrong, H. E.; Horton, E. *Proc. Roy. Soc. Lond. Ser. B. Biol.* 1912, *85*, 109-127.
75. Sumner, J. B. *J. Biol. Chem.* 1926, *69*, 435-441.
76. Sumner, J. B.; Somers, G. F. *Chemistry and Methods of Enzymes*; Academic Press, Inc.: New York, 1947.
77. Gazzola, C.; Blakely, R. L.; Zerner, B. *Can. J. Biochem.* 1973, *51*, 1325-1330.
78. Laskowski, M., Sr. *Cold Spring Harbor Monograph Ser.* 1982, *14*, 1-21.
79. Stryer, L. *Biochemistry*; W. H. Freeman and Co.: San Francisco, CA, 1975.
80. Green, J. R. *Proc. Roy. Soc. Lond. Ser. B. Biol.* 1890, *48*, 370-392.
81. Dakin, H. D. *J. Biol. Chem.* 1910, *8*, 97-104.
82. Fleming, A. *Proc. Roy. Soc. Lond. Ser. B. Biol.* 1922, *93*, 306-317.
83. MacMunn, C. A. *J. Physiol. (Lond.)* 1887, *8*, 57-65.
84. Levy, L. *Z. Physiol. Chem.* 1889, *13*, 309-325.
85. MacMunn, C. A. *Z. Physiol. Chem.* 1889, *13*, 497-499.
86. Hoppe-Seyler, F. *Z. Physiol. Chem.* 1890, *14*, 106-108.
87. Keilen, D. *Proc. Roy. Soc. Lond. Ser. B. Biol.* 1925, *98*, 312-339.
88. Neidleman, S. L. *Biotechnol. Genetic Eng. Rev.* 1987, *5*, 245-267.
89. Henriques, V.; Hansen, C. *Skand. Arch. Physiol.* 1901, *11*, 151-165.
90. Wehmer, C. U.S. Patent 515,033, 1894.
91. Currie, J. N. *J. Biol. Chem.* 1917, *31*, 15.

92. Wells, P. A.; Moyer, A. J.; Stubbs, J. J.; Herrick, H. T.; May, O. E. *Ind. Eng. Chem.* 1937, *29,* 653.
93. Weizmann, C. U.S. Patent 1,315,585, 1919.
94. Allgeier, R. J.; Hildebrandt, F. M. *Adv. Appl. Microbiol.* 1960, *2,* 163.
95. Reed, G. *Food Eng.* 1952, *24,* 105.

Chapter 3

Screening of Hydrolase Libraries Using pH-Shift Reagents: Rapid Evaluation of Enantioselectivity

Francisco Moris-Varas, Laura Hartman, Amit-Shah, and David C. Demirjian[1]

ThermoGen, Inc., 2225 W. Harrison St., Chicago, IL 60612

Abstract: The combination of the pH drop caused by the acid released in an enzymatic hydrolysis and the color change of an indicator-containing solution provides an assay system for hydrolase activity. The main benefit is the use of actual substrates instead of the chromogenic or fluorogenic derivatives typically used for hydrolase screening. The best indicators for most applications are those showing a color transition within the operational pH range (7-8) of most hydrolases, like bromothymol blue and phenol red. The enantioselectivity of lipases and esterases can be evaluated using single isomers under the same conditions and comparing the color turnover for each one. A method is described here for the screening of a lipase library (8 enzymes) against a substrate library (44 single isomers corresponding to 22 racemates) in high-throughput amenable way using 384-well microplates.

Continuous progress in molecular biology, especially DNA cloning and mutagenesis technologies, have facilitated the development of large biocatalyst libraries difficult to predict only a few years ago. [1] As a result, the conventional, low-throughput, methods of screening practiced by organic chemists are unable to provide quick answers to comply with the process development timeline of the pharmaceutical companies. The screening process is the most tedious and time consuming step of the bioprocess implementation, and the lack of tools to shorten it makes biocatalysis a less attractive alternative, with the subsequent loss of opportunities for industrial scale implementation. [2]

This paper describes a new high-throughput amenable screening system for hydrolytic enzymes to evaluate activity and selectivity of this important class of biocatalysts. Hydrolases have been used extensively for asymmetric synthesis of

[1]Contributing author.

bioactive principles and pharmaceutical intermediates. They are increasingly important tools for the manufacturing and development of new pharmaceutical, especially the ones involving chiral centers with defined stereochemistry. [3]

Enzyme Sources

Off-the-shelf Enzymes

By far the fastest and easiest route to finding a new enzyme is to find one that exists from a commercial source. This allows for instant access to the catalyst in sufficient quantities to take the project to the next step. The biggest change over the last few years has been the development of large commercial libraries to enhance and simplify biocatalyst discovery. ThermoGen has been developing enzymes using a methodical plan to discover and develop useful new enzymes for synthetic chemists. As a result, the first set of thermostable enzyme families were discovered and commercialized (esterases, lipases and alcohol dehydrogenases).

Although these commercial libraries are becoming available to the researcher, it is not always possible to find an appropriate enzyme from a commercial source. In this case, a custom screening needs to be performed. This screening can be from a collection of microorganisms of from clone banks that have been generated from these organisms or isolated DNA.

Culture Sources

Screening from culture sources has been succesful in many cases. [4] There are, however, several challenges. First, since the media for different organisms will be diverse, the systematic screening of organism banks becomes more difficult. A general solution is to group like strains together for screening on different types of media and to find a media recipe that can support growth of the organism. Strain redundancy is also another concern. Verification that a particular strain is unique can be accomplished by several methods. However, very small differences in the aminoacid composition of related enzymes can lead to significant differences in enzyme activity. This can be overlooked if using a broader genomic comparison tool to eliminate duplicates.

Gene Banks

By setting up the clone banks in a unified or small set of host organisms (*E. coli, Bacillus,* yeast) only a limited number of propagation methods are required, thus allowing systematic screening methods to be carried out easily. If a cloned enzyme is found of interest, it is easier to scale-up and produce in larger quantities. In addition, cloning is a prerequisite to most types of genetic modification of the gene, including directed evolution. In a clone bank the target gene is often removed from its regulatory elements that can repress expression and helps purify the gene away from isozymes and other competing activities. On the other hand, removing a gene from its regulatory elements can actually turn the gene off instead of on, thus masking the activity. The gene may not express well in the host cloning because of codon usage, nucleic acid structure or lethality. The activity from enzymes that are post-translationally modified may be altered or destroyed. In addition, the actual restriction sites used to construct the gene bank can effect whether any activity is observed or not, since the distance from the expression signals are altered. Finally, for each organism a clone bank is developed from, one needs to screen thousands to tens of thousands of clones for each organism to cover the entire genome of that organism.

Enzyme Screening

High-throughput Screening

In the bioprocess developoment timeline, screening is the limiting for the evaluation of properties of an enzyme library. Figure 1 shows the bioprocess development timeline. It can be divided in three parts: enzyme discovery, enzyme engineering and evolution, and process optimization. If a readily available enzyme is found that can catalyze the desired reaction, the process optimization can start immediately. This generally means a better chance for implementation of the process or at least consideration for manufacturing. Unfortunately, in most cases a novel activity has to be found (enzyme discovery), or there is an existing activity that needs to be improved (enzyme engineering and evolution) or both simultaneously.

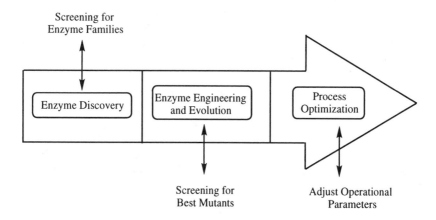

Figure 1. Bioprocess development timeline.

In any of the three segments explained above and shown in Figure 1, the need for powerful assays is apparent. They will allow improved screening of enzyme families for new enzyme discovery, high-throughput screening the best mutants in directed evolution applications, and high throughput adjustment of operational parameters during process optimization.

Table 1 shows the approximate throughput desirable to shorten the above mentioned timeline to a few months (3-6), so the disadvantages confronted by an enzyme-based approach can be overcome and the bioprocess will be given a chance to implementation. The screening application in Table 1 relates with each one of the segments above, and the desired format for the assay is indicated. Although the numbers involved are unmanageable without automation equipment, the fact is that automation alone does not help if the criteria applied are weak or not powerful enough to enable the technology as reliable and dependable.

Table 1. Screening Throughput During Bioprocess Development.

Screening Application	Throughput needed	Desired format
Clones	1,000 / day	Solid phase preferred
Mutants	10,000 / day (including many reaction parameters)	Solid phase preferred, liquid OK
Process Conditions	100-1000 / enzyme (10 pH, 10 temp, 10 cosolvents)	Solid or liquid phase, depending on application

Conventional Screening

The methods typically employed for screening in biocatalysis applications by the organic or synthetic chemists are not high-throughput amenable, since they rely on high-end analytical techniques like HPLC, TLC and GC. On the other hand, three classes of substrate analogs have been used to isolate libraries of different types of carboxylic acid hydrolases (esterases, proteases, lipases, amidases), phosphatases and sulfatases. These substrates consist of carboxylic acid, phosphate or sulfate esters, and amides that contain a detectable probe and a non-detectable moiety with the functionality of interest. [5] The first class of derivatives is the indigogenic substrates. Since these substrates are precipitable they are ideal for first-level hierarchical plate assays since the color develops and stays in the vicinity of the colony. The other two classes of substrates are both soluble substrates that are useful in second-level screens since they are quantitative. Chromogenic substrates based on nitrophenyl or nitroaniline can be used in a quantitative spectrophotometric liquid-assay, but are generally not useful in plate screens since they diffuse readily and are not sensitive enough. Fluorogenic substrates such as those based on umbelliferone or coumarin are at least 1,000 times more sensitive than their chromogenic counterparts. Because of the high sensitivity, they can sometimes be used in plate assays if the reaction is fast enough and can be detected before significant hydrolysis and background appears, but still find their main application in a liquid-phase assay. Some of these substrate derivatives are commercially available, and derivatives resembling substrates of interest can be synthesized to mimic the actual target more closely.

The main limitation of this approach is the presence of the latent colorimetric functionality within the substrate, whose introduction is at least time-consuming and yields a structure for analysis essentially different from the actual target (usually a methyl or ethyl ester). This is especially important when tailoring the activity of a biocatalyst for specific applications, using techniques such as directed or accelerated evolution. Since you usually "get what you screen for", the use of a substrate analog may lead to an improved biocatalyst towards the analog rather than the substrate itself. The possibility of evolving biocatalyst activities for specific applications is extremely important for the industrial success of any biotransformation. This is better understood by comparing biocatalysis and other conventional catalytic approaches: the optimization of the later are intrinsically limited by the nature of the catalyst, while the optimization opportunities of an enzyme-based approach are much broader when customization is taken into account.

Hierarchical Screening

One can gain significant increases in throughput by implementing hierarchical screening approaches. In this type of approach several screening assays are often combined in series to narrow the scope of the screening project step by step. Using this method the easier, but perhaps less accurate, screens are carried out first. The more tedious but quantitative screens are then carried out on only a fractional subset of candidates which have been pre-validated as potentially useful isolates. We have often made use of this type of screening approach because it is rapid, useful, and cost effective. An example 3-level hierarchical screen consists of the following strategy:

Level 1. Most General Screen - Fast and Simple. This type of screen eliminates the majority of negative candidates, and preferentially does not eliminate potential positives.

Level 2. Intermediate Screen(s). This step generally employs more specific substrates or semi-quantitative approaches.

Level 3. Specific Screen. Slowest and most accurate. This type of screen generally employs highly quantitative assays including HPLC, GC, or spectrophotometric or fluorogenic quantitative approaches.

This type of screening approach works because it is subtractive in nature. The first step acts to eliminate the majority of candidates and colonies that do not appear to have any desirable activity at all. As mentioned earlier, the use of substrate analogs almost assures that some potential candidates will be missed which act on a particular substrate of interest, or that some will be found that do not perform on the actual target substrate. For this reason it is important to try and pick substrates that give a good cross-section of activities from the library being screened. In the second step of the screen, expression levels and relative substrate specificity can be determined on a more quantitative basis often using colorimetric candidates like nitrophenyl derivatives. Finally, actual candidate substrates are usually tested to determine which enzymes give the best activity for a particular application.

Actual Substrate Assays

As enzyme libraries and directed evolution applications grow and high throughput screening methods develop, new types of assays which utilize the specific target substrates instead of substrate analogs are needed. The use of a pH-responsive method allows for the rapid screening of hydrolase libraries using the actual substrates rather than chromogenic or fluorogenic derivatives. This is based on the pH drop that happens as the reaction procceds, and the carboxylic acid (from the actual substrate) is released. This drop can be monitored by the change in color of a pH indicator, provided the color profile of the indicator falls into the pH range of the enzymatic activity. Several kinds of reactions have been described including enzyme-catalyzed processes with hexokinase[6-8] and cholinesterase, [9] and in enzyme-free studies of carbon dioxide hydration. [10] Since the 1970's, this strategy has been used in kinetic analysis of enzyme reactions. Examples of this include human carbonic anhydrase, [11] amino acid decarboxylases[12] and serine proteases. [13] The progress of the hydrolysis can be monitored by visual inspection of the solution color after the enzyme has been added or by using a microplate reader to get a quantitative reading. Recently, the use of pH indicators has been extended to monitor the directed evolution of an esterase on a plate assay using a whole cell system, rather than the isolated enzyme. [14]

With regards to the enantioselectivity of hydrolases, the idea is to compare the reactivity (in this case color change) of pairs of enantiomers corresponding to the same racemic mixture and thus estimate the enantioselectivity. A large turnover difference between isomers means there is a good chance of succesful kinetic resolution if the racemic mixture is subjected to the enzyme displaying such time difference. The method involves the use of single isomers, so the kinetics obtained in this way do not reflect the competition that exists when hydrolyzing the racemic mixture, and therefore E value is only approximate. Kazlauskas has solved the problem of competition by using a reference non-chiral additive in classical chromogenic substrate assays. [15] The same author developed a quantitative method (Quick E) for the evaluation of the enantioselectivity (without considering the competition factor) for actual substrates based on a pH indicator/buffer system (p-nitrophenol / BES) with equal pK_a so the linearity of the color transition allows the quantitation of the enantioselectivity. [16]

Despite its impressive accuracy and sensitivity, the method needs special instrumentation (microplate reader) since the color transition cannot be visualized and involves heavy data management. For many applications these can be avoided by using an indicator that effectively turns color so the monitoring could be simplified. In the case described in this paper, the linearity of the assay (and consequently its accuracy) is compromised by the difficulty of choosing a pair buffer / indicator with same pK_a and yielding color change. The bromothymol blue / potassium phosphate system falls within 0.1 units of pK_a (making it suitable for quantitation) allows a nice blue-yellow color transition and uses a common buffer for hydrolase-catalyzed biotransformations. Our goal is to use the method not for the quantitation but for the screening of large amounts of enzymes and / or substrates in high-throughput mode, allowing a quick identification of the highest enantioselectivities, discarding the poor-to-moderate enantioselectivities that will not be acceptable for the development of a biocatalytic resolution. This fits the strategy of a hierarchical screening mentioned above for the identification of the best catalyst at an early step, eliminating the weakest candidates for a more streamlined process-viability study.

Results and Discussion

pH-Shift Reagents

The idea of combining the pH drop of hydrolase-catalyzed reactions with the change of color of an indicator dye associated with a pH transition is depicted in the Figure 2.

$$RCO_2{}^{Et} \xrightarrow[\text{KPi buffer (pK}_a=7.20)]{\text{Hydrolase}} RCO_2{}^- \; + \; H^+$$

$$In^- \; + \; H^+ \xrightarrow{\;\;pH = 6\text{-}8\;\;} InH$$

| blue | Bromothymol Blue (pK$_a$=7.30) | yellow |
| red | Phenol Red (pK$_a$=8.00) | yellow |

Figure 2. Basic principles supporting the pH-shift reagent assay

The use of phosphate buffer at a pH equal to its pK$_a$ (7.20) [17] is a standard choice, compatible with the estability of the majority of lipases and esterases and their activity profile, and also satisfies the need for mild conditions (neutral pH) if sensitive substrates need to be hydrolyzed. It is also usual to choose a pH slightly above or below neutral if the enzyme selected shows better activity under these conditions. [18]

Since pH indicators have been described for most of the spectrum of pH (1-14), other pairs buffer/indicator could be useful in the evaluation of hydrolases with optimal activities at acidic pH (like those obtained from acidophilic organisms) or basic pH (in case the spontaneous substrate hdyrolysis is not significant). The use of enzymes at off-neutral pH values could help multifunctional substrates going into solution, reducing or avoiding the use of cosolvent, which could have a deleterious effect in the reaction rate.

We focused our efforts on the study of Bromothymol Blue (BTB), with Phenol Red as an alternative, since their pK$_a$ values are the closest to that of the buffer we were interested in using and their color transitions show high contrast (as opposed Neutral Red, whose pK$_a$ is also very close but displays poor distinction between red and amber. Preliminary experiments showed that the concentration of indicator dye in the reaction had no effect on the reaction rate, suggesting that the indicator was not acting as an inhibitor of enzyme activity (data not shown). Control experiments using BSA as the protein source caused no change in indicator color and established that pH changes in solution were the result of enzyme catalyzed hydrolysis. Further tests of reaction solutions containing enzymes and indicators without substrates established that color changes in the solutions were not the result of buffer salts or the enzymes themselves.

Substrates and Enzymes

Substrate library
In order to prove the concept of high-throughput enantioselectivity evaluation using this method, we arranged a library of 52 enantiomers corresponding to 26 different substrates. As it can be seen in Figure 3, the 26 substrates are: R and S methyl 2,2-dimethyl-1,3-dioxolane-4-carboxylate (1), R and S methyl 1-methyl-2-oxocyclohexane propionate (2), R and S methyl 2-chloropropionate (3), R and S methyl lactate (4), R and S glycidyl butyrate (5), D and L tryptophan methyl ester (6), R and S methyl mandelate (7), R and S methyl 3-hydroxy-2-methylpropionate (8), R and S methyl 3-hydroxybutyrate (9), R and S ethyl 4-chloro-3-hydroxybutyrate (10), 1R,5S and 1S,5R oxabicyclo[3.3.0]oct-6-en-3-one (11), 1R and 1S menthyl acetate (12), 1R and 1S neomenthyl acetate (13), R and S 3-hydroxy-3-methyl-4,4,4-trichlorobutyric-ß-lactone (14), R and S dimethyl malate (15), D and L dimethyl 2,3-O-isopropylidenetartrate (16), R and S methyl mandelate acetate (17), R and S methyl 3-hydroxy-3-phenylpropionate (18), R and S indanol acetate (19), R and S 1-phenetyl alcohol acetate (20), R and racemic O-Acetyl mandelonitrile (21), R and S acetyl α-Hydroxy-γ-butyrolactone (22), R and S 2-methylglycidyl 4-nitrobenzoate (23), R and S glycidyl 4-nitrobenzoate (24), R and S α-methyl-1-naphtalene-methanol acetate (25), and R and S α-methyl-2-naphtalene-methanol acetate (26).

Every substrate contains at least one chiral center defined as an asymmetric carbon (none of the substrates display heteroatom asymmetry, or chirality consequence of restricted conformation or helical isomers), and some present two (oxabicyclo[3.3.0] oct-6-en-3-one, 11 and dimethyl 2,3-O-isopropyliden tartrate, 16) or even three asymmetric atoms (menthyl acetate 12 and neomenthyl acetate 13). In the case of 11 and 16 the molecules have *cis* configuration, so only one pair of enantiomers (of the two existing pairs) is being tested; and for 14 and 15 the same is true since two out of the three asymmetric centers are preset to test only a pair of stereoisomers. Of course the assay described could be used to evaluate selectivity for all theoretical stereoisomers, but we will limit this study to pairs of isomers for simplicity.

The substrates are commercially available as single isomers from fine chemical companies like Aldrich or Fluka. The assay works in the hydrolysis direction, therefore most substrates can be used as received, but in some cases (substrates 17, 19, 20, 21, 22, 25,26) only the enantiomerically pure alcohol is available, and the corresponding ester (generally acetate) has to be prepared (by acetylation using acetic anhydride or acetyl chloride in pyridine). Some substrates present two ester functions (15, 16, 17, 22), in this case the hydrolysis could take place in either one. It has to be noted that the assay detects overall hydrolysis, but it does not provide information on the regioselectivity of the reaction. All substrates are stocked as acetonitrile solutions of fixed concentration (250 mM) and are dispensed to the reaction in such an amount that the organic solvent concentration does not exceed 7%, thus allowing undisturbed hydrolytic activity to take place.

In the case of mandelonitrile (21), only the R isomer is available and it was compared to the racemic mixture. The comparison is valid if the single isomer (R) is the slow reacting one so the racemic mixture will change color faster.

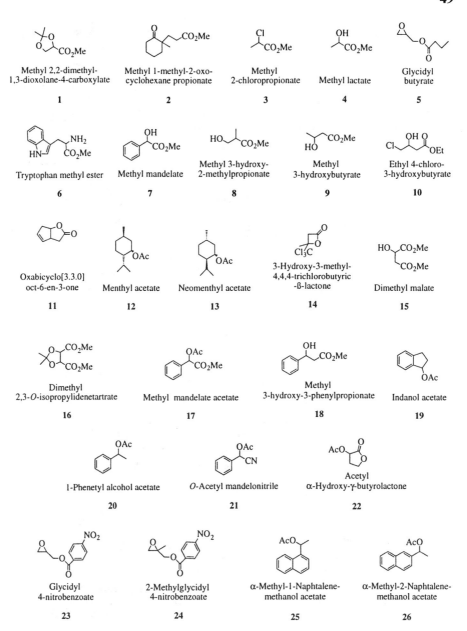

Figure 3. Substrate Library for Hydrolase Enantioselectivity Assay

Enzyme Library

For the initial experiments, we chose the eight lipases offered by Amano Pharmaceutical (Japan), since much information is available on them.[19] The lipases are named A, AK, AY, F, M, N, PS and R. This is an ideal number to show how the assay work, while the high-throughput potential remains apparent. The lipase from *Pseudomonas cepacia* (PS lipase) is well known for enantioselective hydrolysis as well as transesterifications in organic solvents. [20] Some of the substrates shown in Figure 3 (**17-22**) display high enantioselectivity (E) for the development of a successful biocatalytic resolution. Table 2 shows the E values, fast-reacting isomer and the kind of reaction described in the literature. Since many of them are available as alcohols, the E values described are for the transesterification reaction, not for the hydrolysis, but this experiment will also prove valid (in an indirect way) to assess the enantioselective outcome of the reverse reaction.

Table 2.- Best Library Substrates for Amano Lipase PS

Substrate name (number)	E	fast isomer	Reaction	Ref.
methyl mandelate acetate (**17**)	35	S	acetylation	21,22
ethyl 3-hydroxy-3-phenylpropionate (**18**)	127	S	hydrolysis	23
1-indanol acetate (**19**)	412	R	acetylation	24
1-phenylethanol acetate (**20**)	684	R	acetylation	25-32
mandelonitrile acetate (**21**)	26	S	acetylation	33-36
α-hydroxy-γ-butyrolactone acetate (**22**)	high	R	acetylation	37,38

The assay

We have tested the substrates listed in Table 2 with PS lipase and bromothymol blue in order to prove this concept. [39] While in that experiment the assay was run in 96-well plates and 100-200 µL assay volume were being used, in this paper we show the high-throughput potential of the screening by using 384-well plates to screen 4 times as much per microplate with the advantage of using smaller volumes and even lower amounts of enzyme and optically pure substrate. The reactions for substrates **1-22** were set up in a 384-well microplate and the total volume was 80 µL, split as follows: 75µL of Amano lipase from a 10 mg/mL stock solution in 20 mM buffer solution pH = 7.20 containing 0.001% of indicator dye (spun off to avoid turbidity) and 5 µL of acetonitrile substrate solution (from a 250 mM stock in MeCN).

The typical substrate concentration is 15 mM, but for **23-26** the substrates are not soluble under this conditions. For these 4 substrates the use of a detergent like Triton X-100 (0.5%) in the buffer solution allows them to go into solution. Unfortunately, addition of Triton to the stock buffer causes the color to turn yellow, indicating some drop in the pH. Correction of this effect with mild NaOH helps temporarily, but upon standing for a few days the color turns yellow again, indicating some sort of interference of the triton with the indicator / buffer system. For this reason, we do not advise to screen highly apolar substrates using Triton X-100 in this assay. This problem is avoided by using the Kazlauskas assay, which, as explained above, is not visual and needs the help of a microplate reader. The assay buffer in this case is 5 mM BES, containing 0.45 mM PNP and 0.5% Triton X-100. We have used a 96-well microplate reader, so the volumes are 95µL of the assay buffer containing enzyme (10 mg/mL) and 5µL of substrate in MeCN (250mM stock solution). Kazlauskas method

REFERENCES:

1)Demirjian, D. C.; Shah, P.; Morís-Varas, F. *Topics Curr. Chem.* **1998**, *200*, 1.
2)Gerhartz, W. *Enzymes in Industry*; VCH: Weinheim, 1990.
3)Kazlauskas, R. J.; Bornscheuer, U. T. *Biotransformations with Lipases*; VCH: Weinheim, 1999; Vol. 8a.
4)Chartrain, M.; Armstrong, J.; Katz, L.; King, S.; Reddy, J.; Shi, Y. J.; Tschaen, D.; Greasham, R. *Ann N Y Acad Sci* **1996**, *799*, 612-619.
5)Michal, G.; Möllering, H.; Siedel, J. *Chemical Design of Indicator Reactions for the Visible Range*; Bergmeyer, H. U., Ed.; Verlag Chemie: Weinheim, 1983; Vol. I, pp 197.
6)Wajzer, J. *Compt. Rend.* **1949**, *229*, 1270.
7)Darrow, R. A.; Colowick, S. P. *Hexokinase from Baker's Yeast*; Colowick, S. P. and Kaplan, N. O., Ed.; Academic Press: New York, 1962; Vol. V, pp 226.
8)Crane, R. K.; Sols, A. *Animal Tissue Hexokinases*; Colowick, S. P. and Kaplan, N. O., Ed.; Academic Press: New York, 1960; Vol. I, pp 277.
9)Lowry, O. H.; Roberts, N. R.; Wu, M.-L.; Hixon, W. S.; Crawford, E. J. *J. Biol. Chem.* **1954**, *207*, 19.
10)Gibbons, B. H.; Edsall, J. T. *J. Biol. Chem.* **1963**, *238*, 3502.
11)Khallifah, R. G. *J. Biol. Chem.* **1971**, *246*, 2561.
12)Rosenberg, R. M.; Herreid, R. M.; Piazza, G. J.; O'Leary, M. H. *Anal. Biochem.* **1989**, *181*, 59.
13)Whittaker, R. G.; Manthey, M. K.; LeBrocque, D. S.; Hayes, P. J. *Anal. Biochem.* **1994**, *220*, 238.
14)Bornscheuer, U. T.; Altenbuchner, J.; Meyer, H. H. *Biotechnol. Bioeng.* **1998**, *58*, 554.
15)Janes, L. E.; Kazlauskas, R. J. *J. Org. Chem.* **1997**, *62*, 4560.
16)Janes, L. E.; Löwendahl, C.; Kazlauskas, R. *Chem. Eur. J.* **1998**, *4*, 2317.
17)Beynon, R. J.; Easterby, J. S. *Buffer Solutions: The Basics*; IRL Press: Oxford, 1996.
18)Bányai, E. *Indicators*; Pergamon: Oxord, 1971, pp 65-176.
19)Amano technical booklet "Lipases for Resolution and Asymmetric Synthesis".
20)Xie, Z.-F. *Tetrahedron Asymm.* **1991**, *2*, 733.
21)Ebert, C.; Ferluga, G.; Gardossi, L.; Gianferrara, T.; Linda, P. *Tetrahedron Asymm.* **1992**, *3*, 903.
22)Mizayawa, T.; Kurita, S.; Ueji, S.; Yamada, T.; Kuwata, S. *J. Chem. Soc., Perkin Trans. 1* **1992**, 2253.
23)Boaz, N. W. *J. Org. Chem.* **1992**, *57*, 4289.
24)Margolin, A. L.; Fitzpatrick, P. A.; Dubin, P. L.; Klibanov, A. M. *J. Am. Chem. Soc.* **1991**, *113*, 4693.
25)Laumen, K.; Schneider, M. P. *J. Chem. Soc., Chem. Commun.* **1988**, 598.
26)Laumen, K.; Breitgoff, D.; Schneider, M. P. *J. Chem. Soc., Chem. Commun.* **1988**, 1459.
27)Seemayer, R.; Schneider, M. P. *Tetrahedron Asymm.* **1992**, *3*, 827.
28)Bianchi, D.; Cesti, P.; Battistel, E. *J. Org. Chem.* **1988**, *53*, 5531.

54

29)Terao, Y.; Tsuji, K.; Murata, M.; Achiwa, K.; Nishio, T.; Watanabe, N.; Seto, K. *Chem. Pharm. Bull.* **1989**, *37*, 1653.
30)Bianchi, D.; Battistel, E.; Bosetti, A.; Cesti, P.; Fekete, Z. *Tetrahedron Asymm.* **1993**, *4*, 777.
31)Keumi, T.; Hiraoka, Y.; Ban, T.; Takahashi, I.; Kitajima, H. *Chem. Lett.* **1991**, 1989.
32)Gutman, A. L.; Brenner, D.; Boltanski, A. *Tetrahedron Asymm.* **1993**, *4*, 839.
33)van Almsick, A.; Buddrus, J.; Hönicke-Schmidt, P.; Laumen, K.; Schneider, M. P. *J. Chem. Soc., Chem. Commun.* **1989**, 1391.
34)Effenberger, F.; Gutterer, B.; Ziegler, T.; Eckardt, E.; Aichholz, R. *Liebigs Ann. Chem.* **1991**, 47.
35)Inagaki, M.; Hiratake, J.; Nishioka, T.; Oda, J. *J. Am. Chem. Soc.* **1991**, *113*, 9360.
36)Inagaki, M.; Hiratake, J.; Nishioka, T.; Oda, J. *J. Org. Chem.* **1992**, *57*, 5643.
37)Naoyuki, Y.; Miyazawa, K. *Eur. Patent App.*, 1992.
38)Miyazawa, K.; Naouki, Y. *Eur. Patent App.*, 1991.
39)Morís-Varas, F.; Shah, A.; Aikens, J.; Nadkarni, N. P.; Rozzell, J. D.; Demirjian, D. C. *Bioorg. Med. Chem.* **1999**, *7*, 2183.

Chapter 4

Selection of Saprophytic Bacteria and Characterization of Their Fatty Acid Bioconversions during Compost Formation

T. M. Kuo and T. Kaneshiro

Oil Chemical Research, National Center for Agricultural Utilization Research, Agricultural Research Service, U.S. Department of Agriculture, Peoria, IL 61604

A variety of composted materials served as sources of saprophytic bacteria that converted unsaturated fatty acids (UFA) to accumulated products. When oleic acid or 10-ketostearic acid was the selective FA in the bacterial enrichments, *Sphingobacterium thalpophilum*, *Acinetobacter* spp. and *Enterobacter cloacae* represented isolates that produced either hydroxystearic acid, ketostearic acid or incomplete decarboxylations. When ricinoleic (12-hydroxy-9-octadecenoic) acid was the selective UFA, isolates of *E. cloacae* and *Escherichia* sp. produced 12-C and 14-C homologous compounds, and *Pseudomonas aeruginosa* produced a trihydroxyoctadecenoic acid. Various *Enterobacter*, *Pseudomonas,* and *Serratia* spp. appeared to decarboxylate linoleate incompletely. In addition, enrichment cultures from a commercial compost yielded *S. thalpophilum* and *Bacillus cereus* isolates that converted oleic acid to 10-hydroxystearic acid and octadecenamide, respectively. The oleate-selective medium also yielded *Acinetobacter* and coryneform cultures that produced oleyl wax esters. It follows that composts are useful bacterial sources for revealing both the biodegradation and diverse biocatalytic conversion of FAs.

56

Soil bioremediation(*1-3*) and, more generally, compost formation of plant, animal, and petroleum wastes (*4-7*) are vital microbial processes for detoxifying and recycling organic matter. Moreover, compost is the result of concerted microbial activities that are by nature both saprophytic and communal. To retrace a small part of this dynamic process, the fate of lipid substrates can be deduced from recoverable microbes and their reactivity to specific, exogenous fatty acids. A few of the distinctive compost bacteria are identified by species herein, and their degradation and oxidation of long chain unsaturated fatty acids (UFAs) are characterized.

During compost formation, the long chained UFAs presumably are decomposed by FA decarboxylation that shortens hydrocarbons by C-C units from the carboxyl end. In a reversal of anabolic chain elongations, degradation of 18-C FAs may lead to shorter chained ones (16-C to 12-C) and water-soluble acetates. Accumulation of oxygenated FAs in composts (*4, 5*), however, suggest alternative mechanisms to stablize 18-C UFAs. Our results suggest that long chain UFAs are degraded and metabolized mainly by C-C decarboxylations and metabolized by a diverse variety of bacteria. However, distinctive aerobic and facultative anaerobic Gram-negative bacterial strains appear to stabilize UFAs by hydroxylations (*8, 9*) and hydroperoxidation (*10, 11*) bioconversion mechanisms.

Isolation and Identification

Source Materials

Commercial composted manure gave inconsistent data because, among other unknown factors, its origin was not specified (*5*). On the other hand, fresh compost produced under controlled conditions, can give reproducible results. The fresh compost was generated by mixing approximately 40 kg of loose top soil with equal masses of horse manure-sawdust and decaying garden leaves-lawn grass mixtures (*12*). Three 300-mL applications of soybean cooking oil were mixed onto the top layer of this outdoor compost heap at 2-week intervals. Thereafter, soybean oil was added monthly. Compost samples were taken from the top layer after 3, 10, and 12 months and stored in the laboratory in loosely capped jars at room temperature for enrichment culture selections.

Enrichment Culture Selection

An enrichment culture procedure, which is based on the selective advantage of microorganisms to proliferate in a medium containing a specific UFA, was carried out to isolate bioreactive microbes from the composted manure. Oleic acid, linoleic acid, ricinoleic acid, and 10-ketostearic (KSA) were used as selective FAs. In a typical procedure, compost (1.5-2.0 g) and specific UFA (0.5 mL) were added to 100 mL enrichment medium (EM) containing (per L): 5 g glucose, 0.3 g yeast extract, 4.0 g K_2HPO_4, 0.5 g $MgSO_4 \cdot 7H_2O$, 15 mg $FeSO_4 \cdot 7H_2O$ and 1 mL trace minerals; adjusted to pH 7.3 before autoclaving (*5*). Being immiscible with water, linoleic acid (0.5 mL) was premixed with such dispersing agents as 0.1 mL n-octanol, 4 mL potassium stearate suspension (2.5%, w/v), or 1.2 mL hexane. KSA was prepared as a 7.5% substrate (w/v) in 90% ethanol (v/v) and added in 1.35 mL amounts per 50 mL preheated EM. The KSA-containing broth was cooled before adding a compost inoculum (12).

The compost-inoculated broths were incubated aerobically at 200 rpm and 28°C overnight and readjusted to pH slightly above 7.0 with dilute NaOH. The regimen was repeated once or twice after the fourth day by using 2% subtransfer inoculum into fresh UFA-containing EM. Subsequently, each enrichment culture was subjected to dilution-plating onto a selection agar medium. The plates contained EM nutrients supplemented with (per L) 17 g agar, 25 mg bromo-cresol green indicator and 2.0 mL specific UFA (5). Clearly separated colonies were randomly selected for screening. All cultures were maintained on tryptone-glucose-yeast extract (TGY) agar slants and stored at 4°C.

Identification of Bacteria and Their Reaction Products

Screening for active bacteria was carried out in one-step bioconversions in 30 or 50 mL assay broths at 28-30 °C and 180-200 rpm aeration (12, 13). Typically, maintenance cultures were transferred to fresh TGY broth (pH 7.0) and incubated for 1-2 days before transferring 0.3 mL inoculum into 30 mL assay broths in 125-mL Erlenmeyer flasks. The broth contained (per L): 5.0 g yeast extract, 4.0 g glucose, 4.0 g K_2HPO_4, 250 mg $MgSO_4 \cdot 7H_2O$, and 10 mg $FeSO_4 \cdot 7H_2O$ and was adjusted to pH 7.3 before autoclaving (14). The inoculated broth was incubated overnight and readjusted to pH>7.0 before adding 0.3 mL UFA. After 2 to 3 days incubation with FA substrate, the fermentation broth was acidified and extracted twice in one volume of methanol:ethyl acetate (1:9, v/v). The total lipid extracts containing residual FA substrate and accumulated products were concentrated with a rotary evaporator and dried in an evacuated desiccator before weighing.

Portion of the dried lipid extracts was treated with diazomethane and analyzed by GC (Hewlett Packard 5890 Series II Gas chromatograph instrument equipped with a SPB-1 capillary column and coupled to an HP7673 Auto Sampler and Chem Station accessories; Palo Alto, CA) (15). Emergent peaks were identified by retention times relative to standard compounds and area % of the total detected peaks. Typical GC chromatograms of such bioconversion products are shown in Figure 1: KSA, hydroxystearic acid (HSA) and trihydroxyoctadecenoic acid (TOD). Bioconversion products detected by GC were characterized further by an electron impact GC-MS (HP 5890 Gas Chromatograph coupled to an HP-5 capillary column and 70eV HP 5972 Mass Selective Detector) (12). The GC-MS data were compared to those of known chemical structures (12, 13, 15).

Identity of the active cultures was established using an automated Biolog System (Biolog Inc., Hayward, CA) by L. K. Nakamura and H. J. Gasdorf (ARS Culture Collection, Peoria, IL). Several of the bacterial isolates selected from enrichment cultures accumulated either degraded or bioconverted products derived from specific FAs (Table I).

58

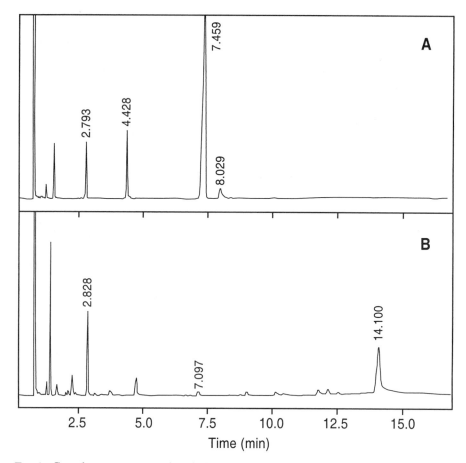

Fig. 1. Gas chromatograms of methyl esters recovered after the conversion of UFAs by compost bacteria. (A) Oleic acid conversion by *Sphingobacterium thalpophilum* strain O22 (NRRL B-23210) to produce 10-ketostearic acid (10-KSA) and 10-hyroxystearic acid (10-HSA). Peaks are internal standard (retention time, RT 2.79 min), oleic acid (RT 4.43 min), 10-KSA (RT 7.46 min), and 10-HSA (RT 8.03 min). (B) Ricinoleic acid conversion by *Pseudomonas aeruginosa* strain Rn30 (NRRL B-23260) to produce a trihydroxyoctadecenoic acid (TOD). Peaks are internal standard (RT 2.83 min), ricinoleic acid (RT 7.10 min), and TOD (RT 14.10 min). Other peaks were not identified.

Table I. Selective Enrichment of Compost Bacterial Isolates Capable of Fatty Acid Conversions (*5, 12, 14*)

Compost Source/ Selective Fatty Acid	Bacterial Isolates	Reaction Products
Fresh compost		
Oleic acid	*Sphigobacterium thalpophilum*; *Acinetobacter* spp.	10-Ketostearic acid (10-KSA); 10-hydroxystearic acid (10-HSA)
10-KSA	*Enterobacter cloacae*	6-Ketotetradecanoate; 4-ketododecanoate
Ricinoleic acid	*E. cloacae; Escherichia* sp.; *Pseudomonas aeruginosa*	12-C and 14-C homologs of ricinoleic acid
	P. aeruginosa	7,10,12-Trihydroxyoctadecenoate
Linoleic acid	*Enterobacter agglomerans*	Hexadecadienoate
Linoleic acid with hexane	*Pseudomonas putida*; *P. viridilivida; Serratia plymuthica*	14-C and 16-C homologs of linoleic acid
Commercial compost		
Oleic acid	*S. thalpophilum; Staphylococcus* sp.; *Flavobacterium gleum*	10-HSA; 10-KSA
	Bacillus cereus	Octadecenamide
	Acinetobacter spp.; coryneform	Oleyl wax esters

Degradation and Bioconversion Products

From oleic acid and 10-KSA

When oleic acid was the selective UFA, the enrichment cultures of fresh compost yielded 12% of the isolates (9 of 74) capable of oleic acid conversion to a 10-HSA and 10-KSA mixture *via* hydroxylation/dehydrogenation reactions. Eight of these 9 bioreactive isolates were identified as *Sphingobacterium thalpophilum* (NRRL B-23206, NRRL B-23208, NRRL B-23209, NRRL B-23210, NRRL B-23211, NRRL B-23212) and *Acinetobacter spp.* (NRRL B-23207, NRRL B-23213). One was unstable and lost its bioreactivity shortly after isolation. Also, 18 unidentified isolates were found to consume or degrade large amounts of oleic acid, leaving less than one half (70-130 mg recovered from 270 mg added per culture) as total extractable lipids. These cultures apparently possessed strong β-oxidation activities to decompose the 18-C FAs to produce CO_2 and short-chained FAs.

In comparison, approximately 8% of the isolates (14 of 165) from enrichment cultures of commercial compost produced a monohydroxy- or monoketofatty acid (*5*). Stable isolates of *S. thalpophilum* (NRRL B-14797) and *Acinetobacter* sp. (NRRL B-14920, NRRL B-14921, NRRL B-14923) converted oleic acid to 10-HSA *via* hydroxylation (*5*) and oleyl wax esters *via* esterification, respectively (*14*). *Flavobacterium gleum* (NRRL B-14798) and *Staphylococcus* sp. (NRRL B-14813) (*5*) exhibited poor bioconversions with oleic acid and lost their viability with repeated subculturing. Also, a stable isolate identified as *Bacillus cereus* (NRRL B-14812) was able to convert oleic acid to octadecenamide *via* an amidation reaction (*5*).

Soil bacteria have been reported to convert oleate to ricinoleate-like compounds (*16*). However, such an activity was not found in our enrichment cultures from compost formation. Other *S. thalpophilum* strains isolated from the compost produced either 10-HSA solely (strain B-14797) or 10-KSA/10-HSA mixtures (Table II). Based on the total amount of bioconversion products and extractable lipids, these *S. thalpophilum* strains fell into four oleate-conversion groups (*17*). Strain B-14797 did not metabolize 10-HSA further (Kuo, unpublished results), nor did other *S. thalpophilum* strains on the accumulated 10-KSA (*18*). Earlier studies with *Bacillus* (*9*), *Corynebacterium* (*19*), *Mycobacterium* (*20*), *Nocardia* (*21*), *Flavobacterium* (*22*), and *Pseudomonas* (*8*), have demonstrated that 10-HSA and 10-KSA production require enzymatic reactions of a hydratase and subsequent secondary alcohol dehydrogenase(s) (*13, 18, 22*). Therefore, strain B-14797 may lack a dehydrogenase enzyme and allows accumulation of 10-HSA only. Because these *S. thalpophilum* strains are stable and easy to maintain, they offer opportunities for the development of a large-scale production of 10-KSA and 10-HSA *via* biocatalytic processes.

With 10-KSA as a selective FA, 14 of 40 random isolates yielded 14-C and 12-C homologs of KSA *via* incomplete decarboxylation reactions (*12*). Three strains possessing this type of bioreactivitiy were identified as *Enterobacter cloacae* A (NRRL

B-23265, NRRL B-23266) and *E. cloacae* B (NRRL B-23264). A reversal to 10-HSA was not detectable in any of the isolates.

Table II. Conversion of Oleic Acid by *Sphingobacterium thalpophilum* Strains Isolated from Compost Cultures

Group	Isolate NRRL No.[a]	10-KSA[b]	10-HSA[b]	Total Product	Yield
		(% total product)		(mg)	(%)
I	B-23206	94	6	126	49
	B-23208	93	7	124	48
II	B-23209	93	7	190	73
	B-23210	94	6	197	76
	B-23211	94	6	185	71
III	B-23212	87	13	80	31
IV	B-14797	0	100	180	69

[a] Strain B-14797 was isolated from commercial compost (*5*) while the others were isolated from fresh compost (*12*).
[b] Abbreviations: 10-KSA, 10-ketostearic acid; 10-HSA, 10-hydroxystearic acid.

From ricinoleic acid

When ricinoleic (12-hydroxy-9-octadecenoic) acid was used as a selective UFA, 20 of 39 random isolates gave detectable bioconversion or decomposition products. Colonies of three isolates gave distinct, clear zones on the selective agar plates. These three isolates were identified as *Pseudomonas aeruginosa* (NRRL B-23256, NRRL B-23258 and NRRL B-23260). Strains NRRL B-23256 and NRRL B-23260 produced a compound with GC-MS profile identical to 7,10,12-trihydroxyoctadecenoate that was produced by strain PR3 (*15*), most likely *via* a hydroperoxidation reaction. Similar to peroxidations of oleate (*11*), the conversion of ricinoleic acid may involve an enzymatic peroxidation-hydroxylation at the C-10 position of 9(*Z*) isomer, a double bond shift from 9(*Z*) to the 8(*E*) position, and a third hydroxylation at the C-7 position (*15*). On the contrary, strain NRRL B-23258 did not yield detectable amounts of the trihydroxyoctadecenoate product. It was concluded subsequently that this type of bioconversion was strain-specific because a number of other *P. aeruginosa* strains examined were unable to convert ricinoleic acid (Kuo and Nakamura, unpublished data).

Lipid extracts from 18 of the 20 positive cultures contained residual ricinoleate substrate as well as homologous 14-C and 12-C compounds (*12*). Small quantities (<4% of the total GC peak areas) of homologous 16-C compound were also detected. Three representatives with this type of decarboxylations were identified as *Enterobacter cloacae* (NRRL B-23257, NRRL B-23267) and *Escherichia* sp. (NRRL B-23259).

From linoleic acid

When linoleic acid was used as the sole source of UFA (36 isolates randomly selected) or as a mixture with n-octanol (33 isolates) or stearate (32 isolates), only 13 out of 101 isolates gave significant amounts of partly decarboxylated compounds identified by GC-MS to be hexadecadienoate (an average of 7% total GC peak areas) (*12*). A representative with this type of bioconversion was identified to be an *Enterobacter agglomerans* biotype 5 (NRRL B-23214). When linoleic acid was suspended in hexane, however, the resulting enrichment cultures produced 21 linoleate-decomposing isolates *via* incomplete decarboxylation. Twelve of these cultures produced 14-C and 16-C homologs. Representatives of these biotypes were identified to be *Pseudomonas putida* type A1 (NRRL B-23263), *Pseudomonas viridilivida* (NRRL B-23261), and *Serratia plymuthica* (NRRL B-23262). In addition, lipid extracts of a few unidentified isolates also contained small amounts of products with GC retention times similar to monohydroxylated and trihydroxylated 18-C fatty acids, presumably *via* undetermined oxygenation reactions. It is noteworthy that all of the isolated bacterial strains originating from composts are saprophytic utilizers of organic waste.

Concluding Remarks

Extensive screenings of microorganisms in soil and water samples from various locations have led to the identification of a *P. aeruginosa* strain that converts oleic acid to a novel dihydroxyoctadecenoic acid (*23*) and a *Clavibacter* sp. that converts linoleic acid to a 12,13,17- trihydroxyoctadecenoic acid (*24*). Notable bioremediation studies have demonstrated the probable enrichment in soils enabling such isolation as *Arthrobacter* strain KCC201 from degrading crude petroleum (*25*), *P. aeruginosa* from gasoline waste (*26*), *Alcaligenes eutrophus* and *Burkholderia* sp. from decomposing 2,4-dichlorophenoxyacetate (*27*), and *Sphingomonas* sp. RA2 from decomposing pentachlorophenol (*1*). However, transient bacterial populations during compost formation related to the utilization and transformation of long chain UFAs have not been well characterized. This paper assesses an enrichment culture method for the isolation of bacteria (Table I) that are capable of UFA degradation and transformation. Our results suggest that decarboxylation, hydroxylation, dehydrogenation, and hydroperoxidation are some alternative mechanisms for stabilizing UFAs during compost formation. Furthermore, the enrichment procedure as described herein offers an effective method for selecting bacteria that specifically transform UFAs to a potentially valuable product.

Acknowledgments

We thank Tamy Leung, Sandra Su, and Richard Hsu for technical assistance, Helen J. Gasdorf for assistance in the identification and maintenance of bacterial cultures that have NRRL accession numbers, Linda Manthey and James Nicholson for GC-MS of samples. We are also grateful to Dr. Lawrence K. Nakamura for culture identification and helpful discussion.

Literature Cited

1. Colores, G. M.; Radehaus, P. M.; Schmidt, S. K. *Appl. Biochem. Biotechnol.* **1995,** *54,* 271-275.
2. Radwan, S. S.; Sorkhoh, N. A.; El-Nemr, I. M.; El-Desouky, A. F. *J. Appl. Microbiol.* **1997,** *83,* 353-358.
3. Head, I. M. *Microbiol.* **1998,** *144,* 599-608.
4. Waksman, S. A. *Humu: Origin, chemical composition, and importance in nature;* The Williams and Wilkins: Baltimore, MD, 1936; pp 191-211.
5. Kaneshiro, T.; Nakamura, L. K.; Bagby, M. O. *Curr. Microbiol.* **1995,** *31,* 62-67.
6. Frostegard, A.; Petersen, S. O.; Baath, E.; Nielsen, T. H. *Appl. Environ. Microbiol.* **1997,** *63,* 2224-2231.
7. Kirchmann, H.; Ewnetu, W. *Biodegradation* **1998,** *9,* 151-156.
8. Wallen, L. L.; Benedict, R. G.; Jackson, R. W. *Arch. Biochem. Biophys.* **1962,** *99,* 249-253.
9. Miura, Y.; Fulco, A. J.. *J. Biol. Chem.* **1974,** *249,* 1880-1888.
10. Gardner, H. W. In *Autoxidation of unsaturated lipids;* Chan, H. W.-S., Ed.; Academic Press: London, 1986, pp 51-93.
11. Guerrero, A.; Casals, I.; Busquets, M.; Leon, Y.; Manresa, A. *Biochim. Biophys. Acta* **1997,** *1347,* 75-81.
12. Kaneshiro, T.; Kuo, T. M.; Nakamura, L. K. *Curr. Microbiol.* **1999,** *38,* 250-255.
13. Kaneshiro, T.; Huang, J. K.; Weisleder, D.; Bagby, M. O. *J. Ind. Microbiol.* **1994,** *13,* 351-355.
14. Kaneshiro, T.; Nakamura, L. K.; Nicholson, J. J.; Bagby, M. O. *Curr. Microbiol.* **1996,** *32,* 336-342.
15. Kuo, T. M.; Manthey, L. K.; Hou, C. T. *J. Am. Oil Chem. Soc.* **1998,** *75,* 875-879.
16. Soda, K. *J. Am. Oil Chem. Soc.* **1987,** 64:1254
17. Kuo, T. M.; Lanser, A. C.; Nakamura, L. K.; Hou, C. T. *Curr. Microbiol.* **1999** (In press).
18. Kuo, T. M.; Lanser, A. C.; Kaneshiro, T.; Hou, C. T. *J. Am. Oil Chem. Soc.* **1999,** *76,* 709-712.

19. Seo, C. W.; Yamada, Y.; Takada, N.; Okada, H. *Agric. Biol. Chem.* **1981,** *45,* 2025-2030.
20. El-Sharkawy, S. H.; Yang, W.; Dostal, L.; Rosazza, J. P. N. *Appl. Environ. Microbiol.* **1992,** *58,* 2116-2122.
21. Koritala, S.; Hosie, L.; Hou, C. T.; Hesseltine, C. W.; Bagby, M. O. *Appl. Microbiol. Biotechnol.* **1989,** *32,* 299-304.
22. Hou, C. T. *Appl. Environ. Microbiol.* **1994,** *60,* 3760-3763.

23. Hou, C. T.; Bagby, M. O. *J. Ind. Microbiol.* **1991**, *7*, 123-130.
24. Hou, C.T. *J. Am. Oil Chem. Soc.* **1996**, *73*, 1359-1362.
25. Borneman, J.; Skroch, P. W.; O'Sullivan, K. M.; Palus, J.; Rumjanek, N. G.; Jansen, J. L.; Nienhuis, J.; Triplett, E. W. *Appl. Environ. Microbiol.* **1996**, *62*, 1935-1943.
26. Foght, J. M.; Westlake, D. W. S.; Johnson, W. M.; Ridgway, H. F. *Microbiol.* **1996**, *142*, 2333-2340.
27. Dunbar, J.; White, S.; Forney, L. *Appl. Environ. Microbiol.* **1997**, *63*, 1326-1331.

Chapter 5

D-Phenylalanine Biosynthesis Using *Escherichia Coli:* Creation of a New Metabolic Pathway

Paul P. Taylor, Nigel J. Grinter, Shelly L. McCarthy, David P. Pantaleone, Jennifer L. Ton, Roberta K. Yoshida, and Ian G. Fotheringham

NSC Technologies, A Unit of Monsanto, 601 E. Kensington Road, Mt. Prospect, IL 60056[1]

D-Amino acids are being increasingly used in the rational design of chiral drugs such as anti-cancer compounds and viral inhibitors. In common with the commercial manufacture of L-amino acids, both chemo-enzymatic resolution approaches and direct single isomer syntheses have been undertaken in the production of D-amino acids and their derivatives. We have developed a fermentation route to the synthesis of D-phenylalanine in *Escherichia coli* K12 using metabolic engineering. The key enzymes used in this system are a DAHP synthase and the chorismate mutase-prephenate dehydratase from *E. coli* in conjunction with an operon of D-amino acid aminotransferase (*Bacillus sphaericus*), L-amino acid deaminase (*Proteus myxofaciens*) and alanine racemase (*Salmonella typhimurium*). These enzymes form a pathway, which deregulates L-phenylalanine synthesis and subsequently converts the L-phenylalanine into its D-isomer. Due to the broad specificity of both the L-amino acid deaminase and the D-amino acid aminotransferase, this system, with slight modifications, can be used to synthesize a wide variety of D-amino acids.

Introduction

In recent years, the demand for D-amino acids by the pharmaceutical industry for the production of single enantiomer and peptidomimetic drugs has shown a worldwide increase. In particular, D-phenylalanine (D-phe) has been in great demand

[1]Current address: NSC Technologies, A Division of Great Lakes Chemical Company, 601 E. Kensington Road, Mt. Prospect, IL 60056

as a chiral synthon for a number of compounds in development. It is difficult and expensive to make D-phe optically pure by chemical synthesis alone, thus production of this amino acid has usually depended on a chemo-resolution approach. D-Phe can also be made using D-aminotransferases and a number of bioconversion approaches have been developed using this technology (*1-4*). Many of these approaches have drawbacks such as high amino donor/amino acceptor costs, equilibrium constraints and sophisticated enzyme reactors. Here I will describe some of our efforts to engineer a strain of *Escherichia coli* K12 (W3110*)* that is capable of producing D-phe by fermentation using glucose as sole carbon and energy source.

D-Phenylalanine by Fermentation (I)

General Considerations

The rationale for our initial attempts to produce D-phe by fermentation using *E. coli* are shown in Figure 1; the two major considerations are the supply of amino acceptor and amino donor. The supply of amino acceptor, phenylpyruvic acid (PPA) can be achieved by the combined effects of deregulation the L-phenylalanine (L-phe) biosynthetic pathway and blocking the terminal step of the pathway. Commitment of carbon to the common aromatic pathway is achieved by relieving the feedback inhibition at the first step of the pathway namely the condensation of erythrose-4-phosphate (E4P) and phosphoenol pyruvate (PEP) by 3-deoxy-D-*arabino*-heptulosonate synthase (DAHP synthase*)*. *E. coli* has three DAHP synthases coded by *aroF*, *aroG* and *aroH*, whereas L-phe and L-tyr tightly regulate the *aroF* and *aroG* gene products respectively, the *aroH* gene product is only 60% feedback inhibited by tryptophan (*5*). Thus overexpression of the *aroH* gene product effectively commits carbon to the common aromatic pathway.

The next major regulation point is at prephenate dehydratase (*pheA*) which is tightly feedback inhibited by L-phe (*6*). To relieve this regulation a cloned *pheA* gene (*pheA*34) which is completely feedback inhibition resistant was added (*7*). Finally, the terminal step in the pathway, PPA to L-phe, needs to be blocked to prevent the synthesis of L-phe and ensure that only the D-enantiomer is synthesized. To block this step effectively, the genes for aspartate aminotransferase (*aspC*), tyrosine aminotransferase (*tyrB*) and branched chain aminotransferase (*ilvE*) need to be disrupted as all three can transaminate PPA to L-phe (*8*).

Supply of amino donor in the form of D-alanine (D-ala) is via host-synthesized L-alanine using the catabolic alanine racemase from *E. coli* (*9*) cloned and overexpressed to make the D-isomer. The host strain also needs to be a D-alanine dehydrogenase (*dadA*) mutant to prevent the catabolism of D-alanine (*10*) and D-phenylalanine (*11, 12*) by this enzyme.

The final step, the transamination of PPA to D-phe, is carried out by the D-aminotransferase (DAT) from *Bacillus sphaericus* (*13*). Plasmid maps showing the construction of the various production plasmids are shown in Figure 2 and the plasmid/strain combinations are shown in Table I.

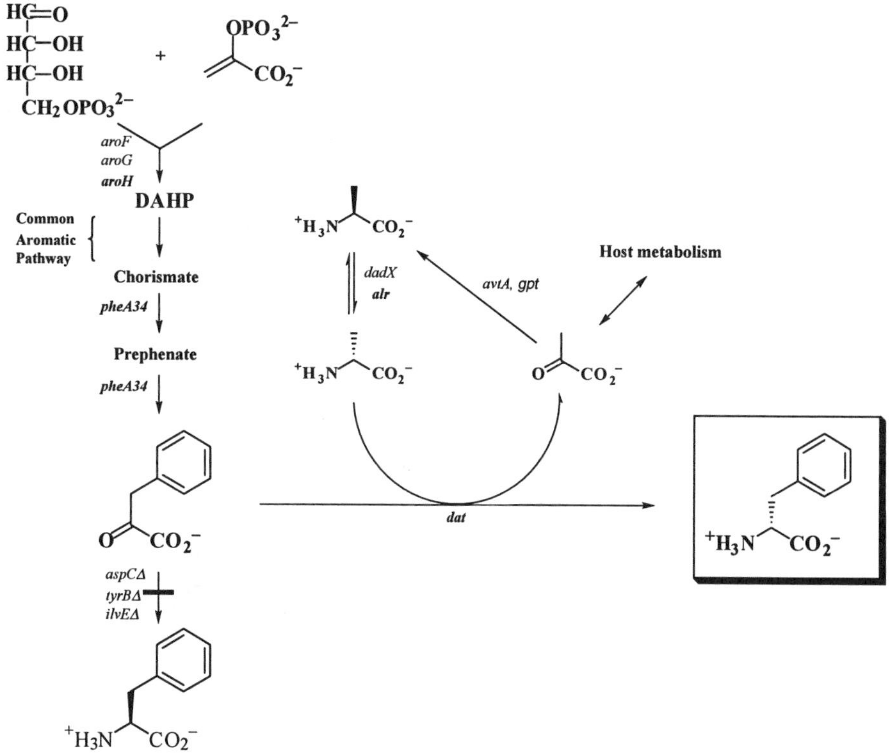

Figure 1. Schematic of the L-phenylalanine pathway which has been deregulated by the addition of a cloned 3-deoxy-D-arabino-heptulosonate-7-phosphate synthase (aroH) and a fully feedback inhibition resistant prephenate dehydratase (pheA). The pathway is blocked at phenylpyruvic acid (PPA) by deletion of the three genes capable of carrying out this reaction, aspartate aminotransferase (aspC), tyrosine aminotransferase (tyrB) and the branched chain aminotransferase (ilvE). Transamination of PPA to D-phe is carried out by the Bacillus sphaericus D-aminotransferase (dat) and amino donor is supplied by the host's alanine biosynthetic pathway. A cloned alanine racemase gene (dadX) from Escherichia coli is present to ensure adequate supply of D-alanine (D-ala).

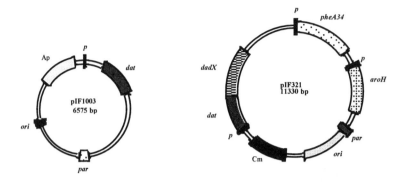

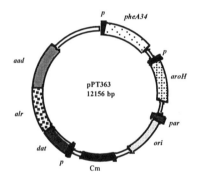

Figure 2. Plasmid constructs used in D-phenylalanine production strains. pIF1003 is a derivative of pBR322(16) carrying the D-aminotransferase gene from Bacillus sphaericus (dat, constituitively expressed from a modified, constitutive pheA promoter) and the partition locus (par) from pLG338(17). pIF321 is a derivative of plG338 carrying the following genes which are constituitively expressed; a fully feedback inhibition resistant prephenate dehydratase (pheA34) from E. coli; a tryptophan repressible 3-deoxy-D-arabino-heptulosonate synthase (aroH) from E. coli; a dat gene from B. sphaericus; the catabolic alanine racemase (dadX) from E. coli and an L-amino acid deaminase (aad) from Proteus myxofaciens. The later three genes are in an operon under control of a modified, constitutive pheA promoter. pPT363 is the same as pIF321 but the dadX gene has been exchanged for the alanine racemase gene (alr) from Salmonella typhimurium.

Table I. Strains/Plasmid Combinations

Host Strain[a]	Genotype	Plasmids	D-Phe Strain
PT100[b]	aspCΔ::kan tyrB::tn10 ilvMEDΔ	pIF321, pIF1003	RY352
RY344[b]	aspCΔ::kan tyrB::tn10 ilvMEDΔ dadAΔ	pIF321, pIF1003	RY347
IF3[c]	aspCΔ::kan dadAΔ	pPT363	IF3/pPT363
IF3[c]	aspCΔ::kan dadAΔ	pPT363, pIF1003	NS3308

[a] All strains were derived from *E. coli* K12, W3110.
[b] Requires asp, phe, tyr, ileu, leu, and val for growth in minimal medium.
[c] Requires no supplements for growth in minimal medium.

Fermentation Using Aromatic L-Aminotransferase Minus Host Strain

Initial attempts to overproduce D-phe were hampered by the finding that strains carrying deletions in *aspC*, *tyrB* and *ilvE* displayed pleotropic growth defects and cell lysis when grown in defined or semi-defined media, even when supplemented with all the required aromatic and branched chain amino acids. Cultures grown in shake flasks in defined medium supplemented with yeast extract did however, show some accumulation of D-phe. A comparison of host strain PT100 and RY344 (Table I) shows that the presence of plasmid pIF1003 and pIF321 does confer synthesis of modest amounts of D-phe (Table II). Further, the addition of a *dadA* lesion significantly increases this synthesis (Table II). Addition of extra amino donor and/or amino acceptor also increased the D-phe synthesis (as might be expected) but again significant accumulation was only seen in strains with a *dadA* lesion, strongly suggesting that D-alanine dehydrogenase is degrading most of the D-phe synthesized.

Attempts to improve the growth of RY347 by media supplementation were met with limited success. Of the many supplements tried only the addition of L-aspartate (5 g/L) improved the growth significantly, presumably because aspartate is a precursor to several essential amino acids required for normal growth which cannot be synthesized due to the aromatic aminotransferase lesions. However, even with the addition of aspartate, high density fermentation was not possible due to cellular lysis. This strongly suggests that the aromatic aminotransferases serve some other, as yet unknown biosynthetic function in *E. coli*.

D-Phenylalanine by Fermentation (II)

Modified D-Phenylalanine Synthetic Pathway: The Addition of L-Amino Acid Deaminase

It is clear from the preceding discussion that a different approach was needed to engineer an *E. coli* strain, which could synthesize D-phe in significant amounts. The alternative approach is shown in Figure 3. Firstly, the host strain was transduced back to *tyrB*[+], *ilvE*[+] thus alleviating the growth problems associated with the triple transaminase mutants and also relieving the requirement for amino acid supplements. Secondly, the *E. coli* catabolic alanine racemase gene was replaced with the biosynthetic alanine racemase (*alr*) from *Salmonella typhimurium* (*14, 15*) in an effort to improve amino donor supply. And finally, the key to this alternative route was the addition of an L-amino acid deaminase (L-AAD) from *Proteus myxofaciens*. This enzyme has a broad substrate range (Table III) and is very active on L-phe, while exhibiting no activity on the amino donor D-ala. The addition of L-AAD allows the DAT to compete with the host's L-aminotransferases for substrate; thus the PPA produced from the deamination reaction is continually transaminated to D-phe. Table IV shows the initial shake flasks results comparing strains IF3/pPT363 and NS3308 (Table I). Significant accumulation of D-phe was detected in both of these strain/plasmid combinations with NS3308 (high DAT) showing the highest yield.

Table II. Shake Flask Results of Aromatic L-Aminotransferase Deleted Strains

Strain	Genotype	Plasmids	No Additions	L-ala[a]	PPA[a]	L-ala + PPA[a]
				D-phe ($\mu g/ml/OD_{600}$)		
PT100	aspC tyrB ilvE	None	0.0	0.0	0.0	0.0
RY352	aspC tyrB ilvE	pIF321, pIF1003	6.3	6.3	25.6	17.9
RY347	aspC tyrB ilvE dadA	pIF321, pIF1003	34.1	41.9	64.7	96.2

[a] 1 g/L of each substrate

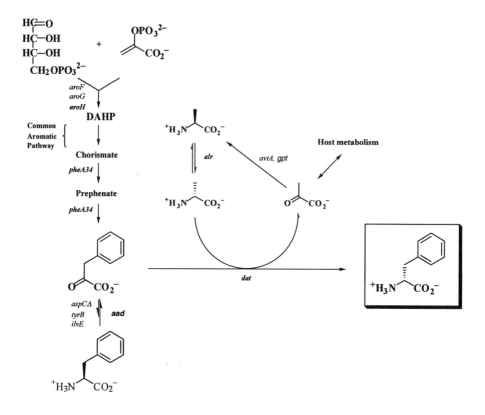

Figure 3. Schematic of alternative route to D-*phenylalanine utilizing* L-*amino acid deaminase (aad) in the presence of tyrosine aminotransferase (tyrB) and branched chain aminotransferase (ilvE). The alr gene from Salmonella typhimurium supplies alanine racemase activity.*

Table III. Substrate Profile of the L-Amino Acid Deaminase from *Proteus myxofaciens*

Amino Acid	R	Activity	Amino Acid	R	Activity
Ala	-CH$_3$	no	Leu	(isobutyl, CH$_3$/CH$_3$)	yes
Arg	(propyl)-N(H)-C(=NH)-NH$_2$	yes	Lys	(butyl)-NH$_2$	yes
Asn	CONH$_2$	yes	Met	-S-CH$_3$	yes
Asp	CO$_2$H	no	Phe	(benzyl)	yes
Cys	SH	yes	Pro	(pyrrolidine, H-N)	no
Glu	CO$_2$H	no	Ser	OH	yes
Gln	CONH$_2$	yes	Thr	CH$_3$/OH	no
Gly	-H	no	Trp	(indole, HN)	yes
His	(imidazole, HN-N)	yes	Tyr	(phenol, OH)	yes
Ile	CH$_3$/CH$_3$	yes	Val	CH$_3$/CH$_3$	yes

Table IV. Shake Flask Results of IF3 Derived Strains

Strain	L-phe (48hr) mg/ml	D-phe (48 hr) mg/ml
IF3/pPT363	0.10	0.49
NS3308	0.08	0.92

Table V. Biosynthesis of D-Phenylalanine by Fermentation using NS3308

L-Phe (g/L)[a]	D/L-Ala (g/L)[a]	D-Phe (g/L)[b]	ee (%)	% Conversion[c]
0	0	1.12	41	n/a
0	20	4.15	100	n/a
20	20	13.70	88	68

[a] 12 hr fermentation followed by 12 hr substrate feed.
[b] Titer after 48 hours.
[c] Percentage conversion of L-phe (fed as substrate) to D-phe.

NS3308 was tested further by fermentation; the results of a series of 12L fermentations where NS3308 was grown with and without combinations of amino acceptor (supplied as L-phe) and amino donor (D-ala) are shown in Table V. Under normal fermentation conditions D-phe was synthesized, but at relatively low titer and the enantiomeric excess (ee) was poor. The addition of an amino donor feed increased the yield of D-phe four fold and significantly improved the enantiomeric purity. This suggests that the reaction is limiting for amino donor; that is, not enough D-ala is being synthesized by the host. When both L-phe and D-ala were fed to the fermentation, 68% of the L-phe added was converted to D-phe although not all of the L-phe was deaminated. This did show however, that the reaction could be run as a fed-fermentation/bioconversion and that this approach could be applicable to making other D-amino acids (see following chapter by D. P. Pantaleone *et al*).

Summary

A novel biosynthetic route for synthesizing D-phe by glucose fermentation was engineered into *E. coli* K12. Although far from optimized, it is clear that this is a valid route to D-phe and we are currently applying the knowledge gained here to proprietary strains that have greater L-phe synthesis capacity. The issue of amino donor supply needs to be addressed and we are currently researching ways to boost the D-ala synthesis in our host strains. This is somewhat complicated by the fact that D-ala is toxic to *E. coli* in concentrations of more than a few grams per liter. Therefore the D-ala synthesis rate needs to closely match that of L-phe synthesis. Due to the broad specificity of both the L-AAD and DAT enzymes, this reaction scheme can also be applied to other commercially important D-amino acids

References

1. Galkin, A.; Kulakova, L.; Yamamoto, H.; Tanizawa, K.; Tanaka, H.; Esaki, N.; Soda, K. *J. Ferment. Bioeng.* **1997**, *83*, 299-300.
2. Galkin, A.; Kulakova, L.; Yoshimura, T.; Soda, K.; Esaki, N. *Appl. Environ. Microbiol.* **1997**, *63*, 4651-4656.
3. Fotheringham, I. G.; Pantaleone, D. P.; Taylor, P. P. *Chim. Oggi* **1997**, *15*, 33-37.
4. Taylor, P. P.; Pantaleone, D. P.; Senkpeil, R. F.; Fotheringham, I. G. *Trends Biotechnol.* **1998**, *16*, 412-418.
5. Pittard, J.; Gibson, F. *Curr. Top. Cell. Regul.* **1970**, *2*, 29-63.
6. Dopheide, T. A. A.; Crewther, P.; Davidson, B. E. *J. Biol. Chem.* **1972**, *247*, 4447-52.
7. Nelms, J.; Edwards, R. M.; Warwick, J.; Fotheringham, I. *Appl. Environ. Microbiol.* **1992**, *58*, 2592-8.
8. Christen, P., ed. *Transaminases.*, ed. P. Christen and D.E. Metzler. 1985, Wiley, New York.
9. Wild, J.; Hennig, J.; Lobocka, M.; Walczak, W.; Klopotowski, T. *MGG, Mol. Gen. Genet.* **1985**, *198*, 315-22.
10. Neidhardt, F. C.; *et al.*, *Escherichia coli and Salmonella: Cellular and Molecular Biology, Second Edition: Two Volumes.* 1996: ASM, Materials Park, Ohio. 3008.
11. Raunio, R. P.; Straus, L. d. A.; Jenkins, W. T. *J. Bacteriol.* **1973**, *115*, 567-73.
12. Olsiewski, P. J.; Kaczorowski, G. J.; Walsh, C. *J. Biol. Chem.* **1980**, *255*, 4487-94.
13. Fotheringham, I. G.; Bledig, S. A.; Taylor, P. P. *J. Bacteriol.* **1998**, *180*, 4319-4323.
14. Wasserman, S. A.; Daub, E.; Grisafi, P.; Botstein, D.; Walsh, C. T. *Biochemistry* **1984**, *23*, 5182-7.
15. Galakatos, N. G.; Daub, E.; Botstein, D.; Walsh, C. T. *Biochemistry* **1986**, *25*, 3255-60.
16. Bolivar, F.; Rodriguez, R. L.; Green, P. J.; Betlach, M. C.; Heyneker, H. L.; Boyer, H. W.; Crosa, J. H.; Falkow, S. *Gene* **1977**, *2*, 95-113.
17. Stoker, N. G.; Fairweather, N. F.; Spratt, B. G. *Gene* **1982**, *18*, 335-41.

Chapter 6

Improvement of Enzyme Character by Gene Shuffling

Kiyoshi Hayashi, Qin Wang, Satoru Nirasawa, Tsuyoshi Simonishi, Motomitsu Kitaoka, and Satoshi Kaneko

Enzyme Applications Laboratory, National Food Research Insitiute, Tsukuba, Ibaraki 305–8642, Japan

Since the modification of the character of enzymes by delibrate engineering is quite difficult, new enzymes which have desirable characteristics are continuously being sought. Since the *de novo* design of enzymes is still far beyond of our current understanding, the most commonly used approach for the discovery of new enzymes is to search for them in natural sources. Nevertheless, gene shuffling provides another method of obtaining new enzymes, in this case by altering the character of natural enzymes. The effectiveness of gene shuffling has been demonstrated for ß-glucosidase, lysozyme and xylanase, although the shuffling of genes tends to disturb the correct folding of the proteins that they code for. However, as shown in the case of aminopeptidase, by introducing mutations in the shuffled gene the correct folding of the target enzyme can be achieved.

Since enzymes are essential components in all living things, they have been extensively studied. Most of enzymes have been classified into 3,704 species based on their characteristics (*1*). However, the enzymes so far utilized in industry are limited to around 30 species (*2*). These enzymes are mostly hydrolyzing enzymes such as amylases, proteinases and lipases. There is plenty of potential for the industrial use of other enzymes if they meet certain criteria.

Searching for useful enzymes possessing desirable characteristics is carried out in many laboratories around the world. Two methods are used to obtain such enzymes; conventional screening and gene manipulation. The single most important factor in using a conventional screening method successfully is to establish a method which, by its very nature, helps to concentrate enzymes with desirable characteristics. This can be rather difficult, if not impossible for many types of enzymes. High productivity of the enzyme by the organism used to be one of the essential factors in conventional screening and is no less important after the enzyme has been cloned.

76

Other than looking for the enzymes which exist in nature, biotechnological methods enable us to modify the character of naturally occurring enzymes. One method is to use site-directed mutagenesis to modify a single amino acid residue among the several hundreds comprising the enzyme molecule. However, in this case, a drastic change in the enzyme character will not usually occur unless the mutation is within the catalytic region of the enzyme (3-5). Another method for DNA manipulation is gene shuffling (6-8). Gene shuffling tends to drastically alter the enzyme conformation and character.

We have successfully shuffled the genes of several enzymes of industrial interest including those of ß-glucosidase, lysozyme, xylanase and aminopeptidase. Active enzymes with altered characters were obtained as described below.

Searching for a Partner Enzyme for Gene Shuffling

For gene shuffling, it is first necessary to identify enzymes which share homology with the the target enzyme. This can easily be achieved by using appropriate internet sites such as http://www.ncbi.nlm.nih.gov/BLAST/. Among the enzymes identified, the partner for the target gene can be selected by considering the desired enzyme character. For example, in the case where an increased heat stability is desired, the selected partner should have higher heat stability than the target enzyme. The degree of amino acid sequence identity that the target enzyme and its selected partner share has also to be considered; higher identity tends to produce shuffled enzymes with activity while less identity tends to produce shuffled products which are insoluble and lack activity.

After identifying a suitable partner enzyme, each enzyme's genes can be transferred to a suitable vector.

Determining Which Regions of the Enzymes Are Suitable for Shuffling

Since sufficient data on the conformation and characteristics of the target and partner enzymes is generally not available, it is rather difficult to predict the characteritics of chimeric enzymes obtained by gene shuffling. A useful first approach is to shuffle several regions of the genes. As correct folding is required for retention of enzyme activity, for successful gene shuffling is essential that the product retains the ability to fold correctly. Genes should be shuffled in such a way so as to minimize the distortion of the translated enzyme and for this reason we normally select shuffling sites within regions of high homology.

As gene shuffling is most easily carried out using restriction enzymes, it is useful to locate restriction enzyme sites common to the target and partner genes. Generally this is not a difficult task, and this is especially true when the identity of the amino acid sequence of the target and partner enzymes is higher than around 70%. In such cases the preferred method of shuffling is to use only restriction enzyme sites since it is very easy to shuffle genes in this way. Among the restriction enzyme sites, those found in highly homologous regions of the amino acid sequence are the best choice for to minimizing distortions of the three dimensional conformation of the resulting chimeric protein product.

Overlapping PCR for Gene Shuffling

In cases where suitable restriction enzyme sites are not available, overlapping PCR (Polymerase Chain Reaction) can be used. This method is able to facilitate the shuffling of genes at any desired position (9-10). Overlapping PCR is illustrated in Figure 1, where three sites are used for DNA shuffling. After deciding on the shuffling sites, two fragments are amplified from each gene at the first PCR cycle. One of the each of the primers used in this first cycle possesses a region of the other enzyme's gene, normally 10 nucleotide base pairs in length. This region works as the matching region in the annealing step at the beginning of the second PCR cycle. Therefore the second cycle of PCR does not require any primers, rather the amplified fragments obtained from the first PCR cycle will be diluted 100 times and then subjected to the second PCR cycle. In this reaction, a very small amount of the shuffled gene–will be constructed. The third step of the process is only to amplify the shuffled gene constructed in the second cycle of PCR. The final PCR products are finally ligated into a expression vector.

Since the replication of the gene is repeated many times in the overlapping PCR process and replication errors may occur, a proof reading DNA polymerase rather than a normal DNA polymerase is recommended (11). Finally, of course, DNA sequencing is required to confirm the sequence of the resulting DNA.

Expression of the Constructed Genes

Expression of the constructed genes can be achieved by normal expression methods. As is the case for other genes, when the transformants possessing the plasmid for the chimeric gene show very low enzyme activity, lowering the cultivation temperature and/or co-expression with molecular chaperons such as GroEL/ES often helps to improve enzyme productivity.

Gene Shuffling of ß-Glucosidase

We have successfully improved the heat stability of ß-glucosidase by gene shuffling (12,13). ß-glucosidases from *Cellvibrio gilvus* (CG) and *Agrobacterium tumefaciens* (AT) belonging to family 3 shared 40% amino acid identity. However, despite this high identity, their enzymatic properties are quite different. The heat stability of CG glucosidase is 41 °C while that of AT glucosidase is 67 °C. No significant trans-glycosylation activity was observed in CG glucosidase while high trans-glycosylation activity is evident in AT glucosidase (14). Even though no 3 dimensional structure is available for any family 3 glycosidase, based on amino acid homology the AT and CG ß-glucosidases are considered to have two domains; the N and C-terminal domains. It has been reported that the catalytic amino acid, aspartate, located in the N-terminal region of CG glucosidase and plays vital role in enzyme activity (15).

To investigate the function of the C-terminal domain, chimeric enzyme genes were constructed, where varying lengths of the C-terminal region of the CG were replaced with the corresponding regions of the AT gene. This was achieved by using restriction enzyme sites as shown in Figure 2. After expressing these shuffled genes in *Escherichia*

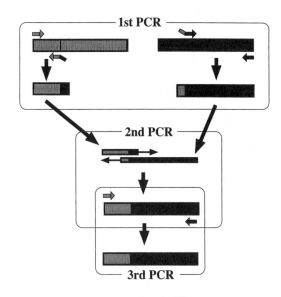

Figure 1. Overlapping PCR for shuffling a gene at any site.

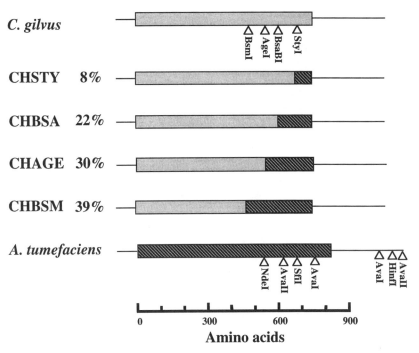

Figure 2. Gene shuffling at the C terminal of β-glucosidase.

coli, the kinetic parameters and the temperature and pH stability of the chimeric enzymes were investigated. The heat stabilities of the chimeric enzymes increased by 6-16 °C with a corresponding increase of AT region from 8- 39 % as shown in Figure 3. The chimeric enzyme which contained 8% of the AT gene was produced in the lowest yield by *E. coli*, this was probably due to a difficulty in protein folding. More recently, five amino acid residues at C terminal end of the CG ß-glucosidase (RGRAR), were found to be quite important in the folding and pH stability of CG ß-glucosidase (*16*).

In order to investigate the role of N terminal domain of CG ß-glucosidase, the AT and CG ß-glucosidase genes were shuffled near the active site in the N terminal region as shown in Figure 4. The shuffled region was selected based on the restriction enzyme sites in the CG gene; the *BamH*I, SacI and *EcoR*I sites. The chimeric genes were constructed in the pET 28a (+) vector and over expressed in *E. coli* BL 21(DE 3), but almost all expressed proteins were produced in an insoluble and inactive form as inclusion bodys.

Refolding Inactive Chimeric ß-Glucosidase

The molecular chaperons, GroEL and GroES have been used in-vivo to solubilize the over expressed proteins (*17*). The co-expression of GroEL/ES with the over-expressed native ß-glucosidase genes (*18*) and a five amino acid deleted mutant of CG (16) was successful in the solubilization of otherwise insoluble protein. However, co-expression of GroEL/ES was not successful in solubilizing the over expressed chimeric ß-glucosidases, even at low growth temperatures; 20-25 °C as shown in Figure 5.

In order to revive the activity of the chimeric enzymes obtained as insoluble proteins, in-vitro refolding of the protein was conducted. Insoluble protein was first purified by using His-tag column chromatography and then solubilized in a solution containing 8 M urea, 5 mM EDTA and 60 mM TrisHCl buffer (pH 8.6). After addition of 5 mM 2-mercaptoethanol, the solubilized protein was reduced by incubating at 40 °C for 2 h. Then, it was oxidized by addition of oxidized glutathione at 80 mg/ml followed by incubation at 40 °C for 30 min. Finally, the reduced and oxidized protein was dialyzed against a buffer containing100 mM TrisHCl (pH 8.0), 1 mM EDTA, 1.6 mg/ml of oxidized glutathione, 0.5 ml/l 2-mercaptoethanol and 8M urea, and the urea concentration was then decreased from 8 M to 0 M over a period of 4 days (*19*).

Using this slow dialysis method both AT and CG ß-glucosidases were successfully refolded into their active forms. However, the chimeric enzymes of Bam-Sac and Bam-Eco showed unstable and very low activity; activity was less than 2% of the native enzymes and varied from batch to batch of their preparations. For these chimeric enzymes it was found that the presence of 2 M urea resulted in increased enzyme activity, suggesting that the preferred enzyme conformation in water alone was different from that of the native enzymes and that the presence of urea enabled a re-orientation of the tertiary structure of the enzymes to give conformations closer to those of the parents.

Another re-folding method with higher reproducibility has been explored for this enzyme system. The same chimeric enzymes (possessing His-Tag peptides) were absorbed onto a His-Trap column (Pharmacia) and then subjected to a urea gradient from 8 M-1 M at a relatively slow flow rate (0.2-0.4 ml/ min, total volume 30-40 column volumes). The folded proteins were then eluted with 0.5 M imidazole.

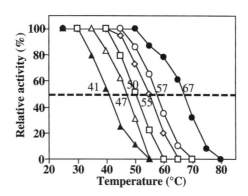

Figure 3. Heat stability of 4 chimeric β-glucosidases and their parent enzymes.

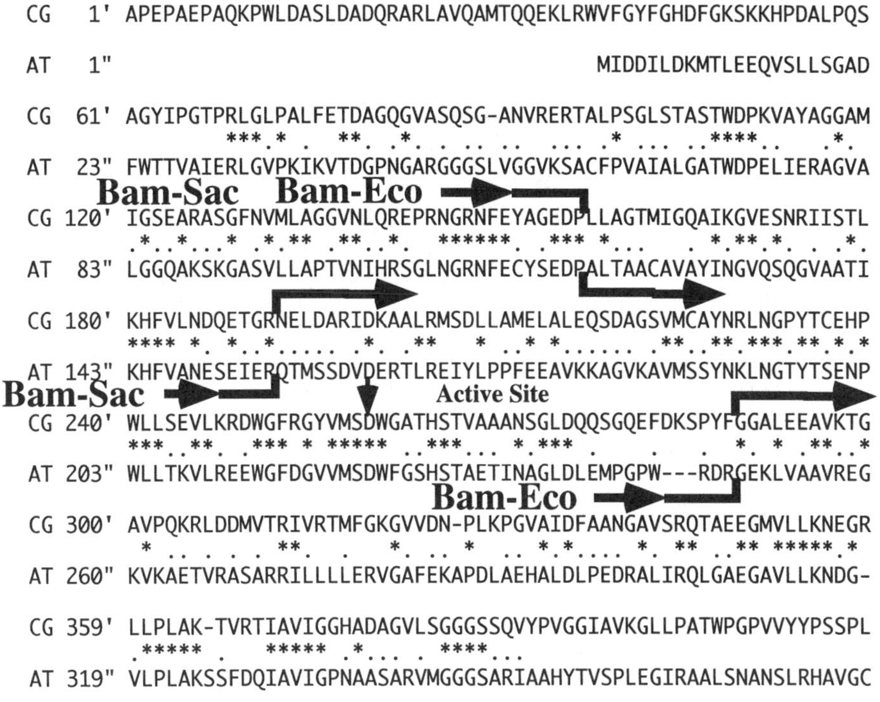

```
CG    1'  APEPAEPAQKPWLDASLDADQRARLAVQAMTQQEKLRWVFGYFGHDFGKSKKHPDALPQS

AT    1"                         MIDDILDKMTLEEQVSLLSGAD

CG   61'  AGYIPGTPRLGLPALFETDAGQGVASQSG-ANVRERTALPSGLSTASTWDPKVAYAGGAM
          ***.*..**..*......... ...*....******..  .*.
AT   23"  FWTTVAIERLGVPKIKVTDGPNGARGGGSLVGGVKSACFPVAIALGATWDPELIERAGVA
```

Bam-Sac   Bam-Eco ➤

```
CG  120'  IGSEARASGFNVMLAGGVNLQREPRNGRNFEYAGEDP LLAGTMIGQAIKGVESNRIISTL
          .*..*..*..*.**..**.*.******.***.*....*.**.*...*.
AT   83"  LGGQAKSKGASVLLAPTVNIHRSGLNGRNFECYSEDRALTAACAVAYINGVQSQGVAATI

CG  180'  KHFVLNDQETGRNELDARIDKAALRMSDLLAMELALEQSDAGSVMCAYNRLNGPYTCEHP
          ****.*..*.*.....*..*.**..*..*..**.**.***.**.*.*
AT  143"  KHFVANESEIERQTMSSDVDERTLREIYLPPFEEAVKKAGVKAVMSSYNKLNGTYTSENP
```

Bam-Sac ➤   **Active Site**   ➤

```
CG  240'  WLLSEVLKRDWGFRGYVMSDWGATHSTVAAANSGLDQQSGQEFDKSPYFGGALEEAVKTG
          ***..**..***.*.***** ..***..*.**..*.*..*..**..*
AT  203"  WLLTKVLREEWGFDGVVMSDWFGSHSTAETINAGLDLEMPGPW---RDRGEKLVAAVREG
```

Bam-Eco ➤

```
CG  300'  AVPQKRLDDMVTRIVRTMFGKGVVDN-PLKPGVAIDFAANGAVSRQTAEEGMVLLKNEGR
          *.....**.*....*..*..*.*.*.....*.**..**.*****.*
AT  260"  KVKAETVRASARRILLLLERVGAFEKAPDLAEHALDLPEDRALIRQLGAEGAVLLKNDG-

CG  359'  LLPLAK-TVRTIAVIGGHADAGVLSGGGSSQVYPVGGIAVKGLLPATWPGPVVYYPSSPL
          .***** . ***** *... . ****...
AT  319"  VLPLAKSSFDQIAVIGPNAASARVMGGGSARIAAHYTVSPLEGIRAALSNANSLRHAVGC
```

Figure 4. Gene shuffling at the N terminal of β-glucosidase.

Although this method gave greater reproducibility than the slow-dialysis method, the end result was the same; the native enzymes were refolded to give active enzymes but none of the chimeric enzymes were refolded successfully.

Gene Shuffling of Microbial Lysozyme

Lysozyme from hen eggwhite is a typical N-acetyl muramidase. Enzymes of similar activity are also produced by microorganisms and plants. This enzyme has the potential to be used as a bio-preservative (20) since it hydrolyses the peptidoglycan in the cell walls of bacteria.

The amino acid sequence of microbial lysozyme produced by *Streptomyces rutgersensis* (21) was subjected to a homology search. It was found that this enzyme showed only 3.3% identity to the lysozyme of hen eggwhite but shared significant homology (50.9% identity) with N,O-diacetyl muramidase of *Streptomyces globisporus* (22). This enzyme has slightly different specificity than lysozyme of hen eggwhite. To identify which regions of these enzymes are primarily responsible for the different substrate specificities, chimeric enzymes were designed based on the alignment of the amino acid sequences as shown in Figure 6. The three shuffling sites were selected in highly homologous regions taking into account the position of the two amino acid residues in the active center.

From three shuffling sites in two enzyme genes, 16 different enzymes can theoretically be produced (including the two original enzymes). By using overlapping PCR, all the possible gene shuffling products were produced (Figure 7). However, five of the new genes were found to contain mutations at several points. These mutations were considered to be generated by repetitive reproduction of DNA during the overlapping PCR. Of the 9 successful chimeric genes, 6 gave chimeric enzymes with activity as shown in Figure 7. The other three gave insoluble proteins as inclusion bodies. Attempts at refolding these aggregated enzymes have not yet produced positive results. When the folding was disturbed by gene shuffling, either aggregated products were generated or small peptides produced by the proteolytic enzyme digestion of the translated products. However, it is quite interesting to note that two third of the chimeric genes gave active enzymes. The characterization of these chimeric enzymes is currently under investigation.

Gene Shuffling of Xylanase

Xylanases are the enzymes that hydrolyse xylan in plant cell walls and they have attracted interest for their use in the bio-bleaching of paper and in the food industry for the production of xylose (23). Based on amino acid sequence, they have been classified into two categories; Family 10 and 11 of the glycosidases (24). *Streptomyces lividans* produces three xylanases; XlnA of family 10, XylB of family 11, and XylC, also of family 11 (25). For gene shuffling we selected xylanases XylB and XylC, both of which hydrolyse by an endo type mechanism to produce xylan oligomers.

In this case the amino acid identity between the two parent enzymes was relatively high (66%) as shown in Figure 8. The two restriction enzyme sites *Sal*I and *Fba*I were

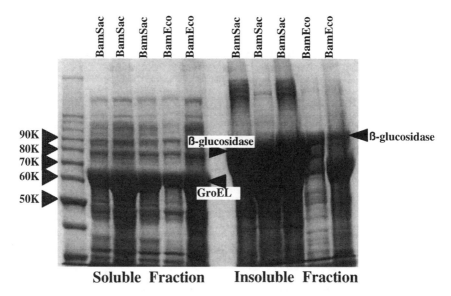

Figure 5. SDS-Page of the crude chimeric β-glucosidases shuffled in the N-terminal region.

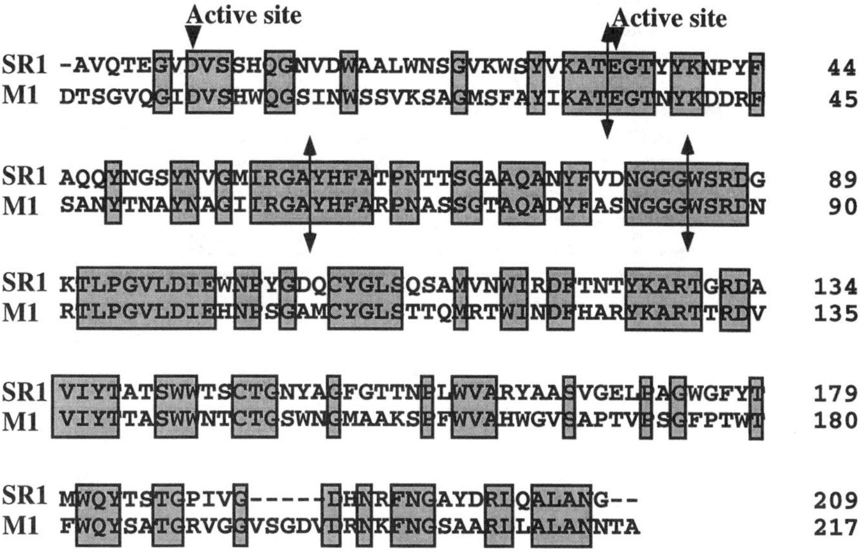

```
                  Active site                    Active site
SR1  -AVQTEGVDVSSHQGNVDWAALWNSGVKWSYVKATEGTYYKNPYF    44
M1   DTSGVQGIDVSHWQGSINWSSVKSAGMSFAYIKATEGTNYKDDRF    45


SR1  AQQYNGSYNVGMIRGAYHFATPNTTSGAAQANYFVDNGGGWSRDG     89
M1   SANYTNAYNAGIIRGAYHFARPNASSGTAQADYFASNGGGWSRDN     90


SR1  KTLPGVLDIEWNPYGDQCYGLSQSAMVNWIRDFTNTYKARTGRDA    134
M1   RTLPGVLDIEHNPSGAMCYGLSTTQMRTWINDFHARYKARTTRDV    135


SR1  VIYTATSWWTSCTGNYAGFGTTNPLWVARYAASVGELEAGWGFYT    179
M1   VIYTTASWWNTCTGSWNGMAAKSPFWVAHWGVSAPTVESGFPTWT    180


SR1  MWQYTSTGPIVG-----DHNRFNGAYDRIQALANG--          209
M1   FWQYSATGRVGGVSGDVDRNKFNGSAARLLALANNTA          217
```

Figure 6. Amino acid alignment of the microbial lysozymes SR-1 and M1.

84

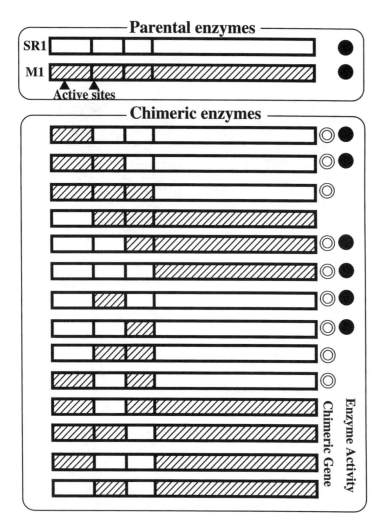

Figure 7. Diagramatic representation of all possible chimeric enzymes that can be produced by shuffling two enzymes by using three shuffling sites, and their enzyme activity.

Of the 14 chimeric genes constructed, only 9 genes were constructed without unexpected mutations. Seven chimeric genes produced enzymes with activity.

XlnB   `MNLLVQPRRRRGPVTLLVRSAWAVALARSP-LMLPGTAQADTVVTTNQEGTN` 52'
           `.*** .. *.******.*.*.*****.**`
XlnC   `MQQDGTQQDRIKQSPAPLNGMSRRGFLGGAGTLALATASGLLLPGTAHAATTITTNQTGT-` 60"

XlnB   `NGYYYSFWTDSQGTVSMNMGSGGQYSTSWRNTGNFVAGKGWANGGRRTVQYSGSFNPSGNA` 113'
           `.* ********. *.*...* * *** .* * ********** ..*. .* .* *** **.`
XlnC   `DGMYYSFWTDGGGSVSMTLNGGGSYSTQWTNCGNFVAGKGWSTGD-GNVRYNGYFNPVGNG` 120"

▼*Sal* I
XlnB   `YLALYGWTSNPLVEYYIVDNWGTYRPTGEYKGTVTSDGGTYDIYKTTRVNKPSVEGTRTFD` 174'
           `* *******************. .***** .*** * *******.**.`
XlnC   `YGCLYGWTSNPLVEYYIVDNWGSYRPTGTYKGTVSSDGGTYDIYQTTRYNAPSVEGTKTFQ` 181"

▼*Fba* I
XlnB   `QYWSVRQSKRT--GGTITTGNHFDAWARAGMPLGNFSYYMIMATEGYQSSGTSSINVGG` 231'
           `********* . ******************* .*.*.********.*.*.*.*`
XlnC   `QYWSVRQSKVTSGSGTITTGNHFDAWARAGMNMGQFRYYMIMATEGYQSSGSSNITVSG` 240"

Figure 8. Amino acid alignment of the two xylanases.

selected for gene shuffling because they were in highly homologous regions and existed uniquely at analogous sites in the two genes.

The chimeric gene BSC (Figure 9) was constructed as follows: Each parental gene was first cloned into the pQE60 expression vector. This gave the plasmids pQEXlnB (from XylB) and pQEXnlC (from XylC). To obtain the insert fragments, sites suitable for binding appropriate PCR primers were located outside the required boundaries of the genes by the aid of appropriate software. The PCR products obtained by using pQEXnlC as the template were then digested by *Sal*I and *Hind*III to yield an insert having cohesive ends. The vector of pQEXlnB was similarly digested by *Sal*I and *Hind*III and the desired DNA fragment which included the bulk of the vector was purified using agarose gel electrophoresis. Finally both the digested vector and PCR fragment were ligated and transformed into *E. coli*. The other chimeric genes depicted in Figure 9 were similarly constructed. It can be seen from this example that by using restriction enzyme sites for gene shuffling, construction of chimeric genes can be quite a simple process.

It was found that all the newly constructed genes gave active enzymes which varied in their heat stability. The heat stability of the parental enzymes was 49 °C for XlnB and 65 °C for XlnC. For the chimeric enzymes BSC, BFC, CSB and CFB, heat stability was 61 °C, 55 °C, 46 °C and 52 °C, respectively. The chimeric enzymes BSC and BFC showed higher heat stability than their dominant parental enzyme, XlnB.

Gene Shuffling of Aminopeptidase in Combination with Random Mutagenesis

Proteolytic enzymes are used extensively in industry and are separated into two main types; endo and exo. The exo type is further classified into the aminopeptidase and carboxypeptidase categories. Among these enzymes, we noticed that there is a relative scarcity of aminopeptidases. We have previously obtained a unique aminopeptidase possessing debittering activity by using traditional screening methods (26,27). The debittering activity of this enzyme was mainly due to its ability to cleave hydrophobic amino acids such as leucine and phenylalanine from the N-terminal ends of bitter peptides (28). This debittering effect is due to the fact that it is the hydrophobic amino acids which are largely responsible for the bitter taste of some peptide mixtures. Another unique feature of this aminopeptidase is that its pro-region functions as an intramolecular chaperone (29).

We have tried to improve the heat stability of this unique aminopeptidase produced by *Aeromonas caviae* by gene shuffling with an aminopeptidase from *Vibrio proteolitica* (30). As a first approach, three shuffling sites were selected and overlapping PCR was employed to construct chimeric genes. However, even though there was a high degree of identity (56.7 %) in the amino acid sequence of the two enzymes, all three chimeric genes were expressed in insoluble form. The refolding of these insoluble proteins using the methods described in the above ß-glucosidase section was not successful.

In order to overcome this problem, mutation of the shuffled genes was attempted. Mutations were introduced by two methods; site-directed mutations were introduced by modifying the synthesized primers used in the first cycle of the overlapping PCR, and random mutations were introduced by error prone PCR (31,32). As a result, an active mutated chimeric gene was produced as shown in Figure 10. Similar to the non-mutated

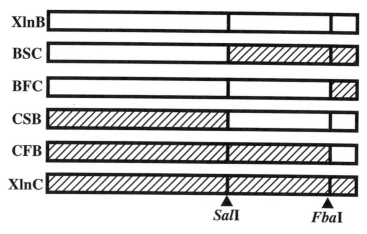

Figure 9. Chimeric xylanases constructed by using the two restriction enzyme sites SalI and FbaI.

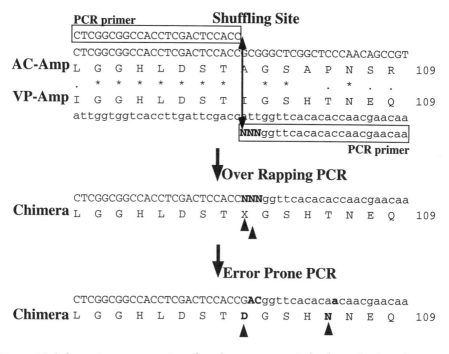

Figure 10. Schematic representation of random mutagenesis for the activation of inactive chimeric aminopeptidase.

Boxes show the nucleotide sequence of the chimeric enzyme around the shuffling site. Bold letters indicate mutated bases and amino acids.

88

chimeric enzymes previously described in this article, it was found that the heat stability of the mutated chimeric enzyme (65 °C) fell between that of the two parental enzymes (50 °C for *A.caviae* and 75 °C for *V. proteolitica* aminopeptidase) as shown in Figure 11. No significant differences in the kinetic parameters *Km* and *kcat* for these enzymes were observed.

This result suggests that correct folding can be achieved in otherwise incorrectly folded enzymes produced by gene shuffling by introducing mutations. Even though the gene was shuffled using sites at highly homologous regions, gene shuffling may have introduced certain torsions which left the protein with a limited ability to form an active three dimensional structure. Random mutagenesis combined with selective screening was a successful method for overcoming such a limitation in the shuffled enzyme.

Conclusion

One of the big advantages of applying the gene shuffling technique in order to improve enzyme character is that it is not necessary to know the three dimensional structure. Although it is desirable to know the enzymes three dimensional structure for designing the shuffling plan and to be able to fully interpret result, as long as the amino acid sequence of the target enzyme is available, it is not difficult to shuffle the gene of an enzyme.

The other advantages of shuffling genes is that the character of the shuffled products can be predicted to some extent. In general, the character of chimeric enzymes constructed by the gene shuffling becomes a mixture of the parental enzymes. This can be quite an advantage especially as the prediction of enzyme character based on the three dimensional structure is almost impossible at present. This method can be regarded as a 'molecular breeding method' for the improvement of enzyme characteristics. The method for gene shuffling itself is not complicated at all. Once the shuffling sites have been selected the shuffled genes can be obtained within several days. However, it is still difficult to determine which shuffling sites to select to obtain enzymes of desired character. As the accumulation of knowledge in this area grows, the outcomes of gene shuffling will become easier to predict, and gene shuffling will provide a relatively accessible way for altering enzyme character

Acknowledgements

This work was supported in part by a grant from the Program for Promotion of Basic Research Activities for Innovative Biosciences.

References

1. Bairoch, A. *Nucleic Acids Res.* **1999**, *27*, 310-311.
2. *Food Enzymes*; Wong D. W. S., Chapman & Hall, New York, NY, 1995; pp 1-16.
3. Damborsky, J.; Bohac, M.; Prokop, M.; Kuty, M.; Koca, *J. Protein Eng.* **1998**, *11*, 901-907.

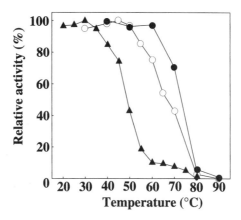

Figure 11. Thermostability of parental aminopeptidases and the activated chimeric aminopeptidase.

Each enzyme was treated at different temperatures for 30 min at pH 8.5 and the residual activities were measured. ▲, aminopeptidase of *A. caviae*; ●, *V. proteolitica*; ○, activated chimeric enzyme.

4. Melnikov, A.; Youngman, P. J. *Nucleic Acids Res.* **1999**, *27*, 1056-1062.
5. Buchholz, F.; Angrand, P. O.; Stewart, A. F. *Nat. Biotechnol.* **1998**, *16*, 657-662.
6. Stemmer, W. P. *Nature* **1994**, *370*, 389-391.
7. Crameri, A.; Raillard, S. A.; Bermudez, E.; Stemmer, W. P. *Nature* **1998**, *391*, 288-391.
8. Harayama, S. *Trends Biotechnol.* **1998**, *16*, 76-82.
9. *Genetic Engineering with PCR*; Horton R. M.; Tait R. C., Ed.: Horizon Scientific Press: Norfolk, 1998; pp 97-106.
10. Zhong, D.; Bajaj, S. P. *Biotechniques* **1993**, *15*, 874-878.
11. Flaman, J. M.; Frebourg, T.; Moreau, V.; Charbonnier, F.; Martin, C.; Ishioka, C.; Friend, S. H.; Iggo, R. *Nucleic Acids Res.* **1994**, *22*, 3259-3260.
12. Singh, A.; Hayashi, K. *J. Biol. Chem.* **1995**, *270*, 21928-21933.
13. Singh, A.; Hayashi, K.; Hoa, T. T. *Biochem. J.* **1995**, *305*, 715-719.
14. Watt, D. K.; Ono, H.; Hayashi, K. *Biochem. Biophys. Acta*, **1998**, *1385*, 78-88.
15. Kashiwagi, Y.; Aoyagi, C.; Sasaki, T.; Taniguchi, H. *J. Ferment. Bioeng.* **1993**, *75*, 159-165.
16. Kim, J. D.; Singh, S.; Machida, S.; Chika, Y.; Kawata, Y.; Hayashi, K. *J. Ferment. Bioeng.* **1998**, *85*, 433-435.
17. Hoshino, M.; Kawata, Y.; Goto, Y. C. *J. Mol. Biol.* **1996**, *262*, 575-587.
18. Machida, S.; Yu, Y.; Singh, S.; Kim, J. D.; Hayashi, K.; Kawata, Y. *FEMS Microbiol Lett.* **1998**, *159*, 41-46.
19. Maeda, Y.; Yamada, H.; Ueda, T.; Imoto, T. *Protein Eng.* **1996**, *9*, 461-465.
20. Hayashi, K.; Kasumi, T.; Haraguchi, K.; Kubo, N.; Tsumura, N. *Agric. Biol. Chem.* **1989**, *53*, 3173-3177.
21. Shimonishi, T,; Nirasawa S.; Hayashi, K. *J. Ferment. Bioeng.* **1999**, *88*, 362-367.
22. Lichenstein, H. S.; Hastings, A. E.; Langley, K.E.; Mendiaz, E. A.; Rohde, M. F.; Elmore, R.; Zukowski, M. M. *Gene* **1990**, *88*, 81-86.
23. Coughlan, M. P.; Hazlewood, G. P. Biotechnol. *Appl. Biochem.* **1993**, *17*, 259-289.
24. Henrissat, B.; Bairoch, A. *Biochem. J.* **1993**, *293*, 781-788.
25. Shareck, F.; Roy, C.; Yaguchi, M.; Morosoli, R.; Kluepfel, D. *Gene*, **1991**, *107*, 75-82.
26. Izawa N.; Hayashi, K. *J. Ferment. Bioeng.* **1996**, *82*, 544-548.
27. Izawa N.; Hayashi, K. *J. Agric. Food Chem.* **1997**, *45*, 543-546.
28. Izawa N.; Hayashi, K. *J. Agric. Food Chem.* **1997**, *45*, 4897-4902.
29. Nirasawa, S.; Nakajima, Y.; Zhang, Z. Z.; Yoshida, M.; Hayashi, K. *Biochem. J.* **1999**, *341*, 25-31.
30. Guenet, C.; Lapage, P.; Harris, B. A. *J. Biol. Chem.* **1992**, *267*, 8390-8395.
31. Leung D. W.; Chen E.; Goeddel D. *Techique*, **1989**, *1*, 11-15.
32. Cadwell R. C.; Joyce G. F. *PCR methods; applications,* **1992**, *2*, 28-33.

Applications: Specialty Chemicals

Chapter 7

Bioconversion of Unsaturated Fatty Acids to Value-Added Products

Ching T. Hou

Oil Chemical Research, NCAUR, Agricultural Research Service, U.S. Department of Agriculture, 1815 North University Street, Peoria, IL 61604 (Telephone: (309) 681–6263; FAX: (309) 681–6340; email: houct@mail.ncaur.usda.gov)

Hydroxy fatty acids are important industrial materials. The hydroxy group gives a fatty acid special properties, such as higher viscosity and reactivity. Presently, imported castor oil is the only commercial source of industrial hydroxy fatty acids. Production of hydroxy (including keto) fatty acids through bioconversion has drawn much attention recently. There are three types of hydroxy fatty acids produced by bioconversion: mono, di, and trihydroxy fatty acids. Chemical structure of trihydroxy unsaturated fatty acid resembles those of plant self-defense substances.

Surplus vegetable oils represent attractive renewable feedstocks for the production of useful chemicals. Unsaturated fatty acids are major components of soybean oil and corn oil, representing 85 and 86%, respectively, of the fatty acids present in the oils. We are investigating microbial conversion of vegetable oils and their component fatty acids to value-added products. One class being hydroxy fatty acid.

The hydroxy group gives a fatty acid special properties, such as higher viscosity and reactivity compared with other fatty acids. Because of their chemical attributes, hydroxy fatty acids are used in a wide range of products, including resins, waxes, nylons, plastics, corrosion inhibitors, cosmetics, and coatings. Furthermore, they are used in grease formulations for high-performance military and industrial equipment. Plant systems are known to produce hydroxy fatty acids. Presently, imported castor oil and its derivatives are the only commercial source of these industrial hydroxy fatty acids. Ricinoleic and sebacic acids, two castor oil derivatives, are classified by the Department of Defense as strategic and critical materials (1). Because of fluctuating supplies and prices for castor oil, some companies have sought alternative raw materials, primarily petroleum-based feedstocks.

Microbial enzyme systems can biotranform unsaturated fatty acids to three types of hydroxy fatty acid products, namely: monohydroxy, dihydroxy, and trihydroxy fatty acids. These products have potential industrial applications.

I. Monohydroxy Fatty Acid

Microbial hydration of unsaturated fatty acid was first reported by Wallen *et al.* (2) who found that a *Pseudomonad* isolated from fatty material hydrated oleic acid at the *cis* 9 double bond to produce 10-hydroxystearic acid (10-HAS) in 14% yield. The 10-HSA was optically active (3,4) and had the D-configuration (4). Seo *et al.* (5) isolated a culture, *Corynebacterium* sp. S-401, from soil, which stereospecifically hydrated oleic acid to 10-ketostearic (10-KSA) and (-)-10R-hydroxystearic acid in 22.4 and 9.1% yield, respectively.

Cells of *Rhodococcus rhodochrous* also hydrated oleic acid to 10-HSA and 10-KSA in 55% and 12% yields, respectively (6). Hydration of oleic acid to 10-HSA was also demonstrated in resting cell suspensions of seven Nocardia species under anaerobic condition (7). *Nocardia cholesterolicum* NRRL 5769 gave a yield exceeding 90% with optimum conditions at pH 6.5 and 40 C. A minor product, 10-KSA was detected. The reaction proceeds via hydration of the double bond as shown by labeling experiments using deuterium oxide and ^{18}O-labeled water. The system was specific for fatty acids with *cis* unsaturation at the 9 position. Anaerobiosis favors bioconversion to 10-HSA (8) while higher pH favors bioconversion to 10-KSA (7).

To date, the microbial hydration of oleic acid was found in *Pseudomonas* (2), *Nocardia* (*Rhodococcus*) (6,7), *Corynebacterium* (5), *Sphingobacterium* (9) and *Micrococcus* (10). Works of El-Sharkawy *et al.* (11) considerably extended the genera of microorganisms known to hydrate oleic acid to include a range of eucaryotic organisms. Strains from several genera including *Absida*, *Aspergillus*, *Candida*, *Mycobacterium*, and *Schizosaccharomyces* were also found capable of catalyzing the hydration of oleic acid.

The stereospecificity of microbial hydrations of oleic acid to 10-HSA were investigated by Yang *et al.* (12) based on the ^{1}H-nuclear magnetic resonance spectral analysis of diastereomeric S-(+)-O-acetylmandelate esters of hydroxystearates (11). They found that while *R. rhodochrous* ATCC 12674-mediated hydration of oleic acid gave an enantiomer mixture 10(R)-hydroxystearic acid and 10(S)-hydroxystearic acid, *Pseudomonas* sp. NRRL B-3266 produced optically pure 10(R)-hydroxystearic acid. The remaining microorganisms investigated (9) hydrated oleic acid to 10(R)-hydroxystearic acid that contained 2 to 18% of 10(S)-hydroxystearic acid enantiomer.

Lanser (13) reported the conversion of oleic acid to 10-ketostearic acid by microorganism from *Staphylococcus* sp. The yield was greater than 90% with less than 5% of co-product, 10-hydroxystearic acid.

Hou reported that *Flavobacterium* sp. DS5 (14) converted oleic acid to 10-KSA in 85% yield. Optimum time, pH, and temperature for the production of 10-KSA were 36hr, 7.5, and 30 C, respectively. Evidence obtained strongly suggested that oleic acid was converted to 10-KSA *via* 10-HSA. Stereochemistry of product 10-HSA produced by strain DS5, determined by ^{1}H-NMR of the mandelate esters of methyl-10-hydroxystearate, showed the 10(R) form in 66% enantiomeric excess.

The *Flavobacterium* DS5 enzyme system also catalyzes the conversion of linoleic acid to hydroxy derivatives. In contrast to oleic acid as substrate which yielded mainly a keto product, linoleic acid as substrate yielded mainly 10-hydroxy-12(Z)-octadecenoic acid (10-HOA) in 55% yield (15).

Strain DS5 hydrated unsaturated but not saturated fatty acids. The relative fatty acid reactivity were in the following order: oleic > palmitoleic > arachidonic > linoleic > linolenic > gamma-linolenic > myristoleic acids.

Strain DS5 also converted α-linolenic acid to 10-hydroxy-12,15-octadecadienoic acid and γ-linolenic acid to 10-hydroxy-6(Z),12(Z)-octadecadienoic acid (15). The enzyme hydrated 9-unsaturation but did not alter the original 6,12, or 15-unsaturations. It is interesting to find that all unsaturated fatty acids tested are hydrated at the 9,10 positions with the oxygen functionality at C-10 despite their varying degree and positions of unsaturations in the fatty acids. (Figure 1).

II. DIHYDROXY UNSATURATED FATTY ACID

Recently, we isolated a bacterial strain (PR3), which converted oleic acid to 7,10-dihydroxy-8(E)-octadecenoic acid (DOD) a process that involves both isomerization and hydroxylation (16,17). Strain PR3, isolated from a water sample at a pig farm in Morton, IL, formed a smooth, round, white colony on agar plate. The microorganisms were motile, short rod-shaped bacteria. Flagella stain showed multiple polar flagellae. Strain PR3 grew aerobically but could not grow anaerobically, and oxidase activity of the cells was positive. Strain PR3 was subsequently identified as a strain of *Psedomonas aeruginosa*. (18).

Chemical structure of DOD was determined by GC/MS, FTIR, and NMR (16). The production of DOD from oleic acid reached a maximum after 48 hr of incubation with a yield of 63%. The yield was later improved to greater than 80% by modifying the culture medium and reaction parameters (19) but continued incubation reduced DOD content in the medium indicating that strain PR3 metabolizes DOD. The production of DOD at >90% was also demonstrated using a cell-free enzyme preparation.

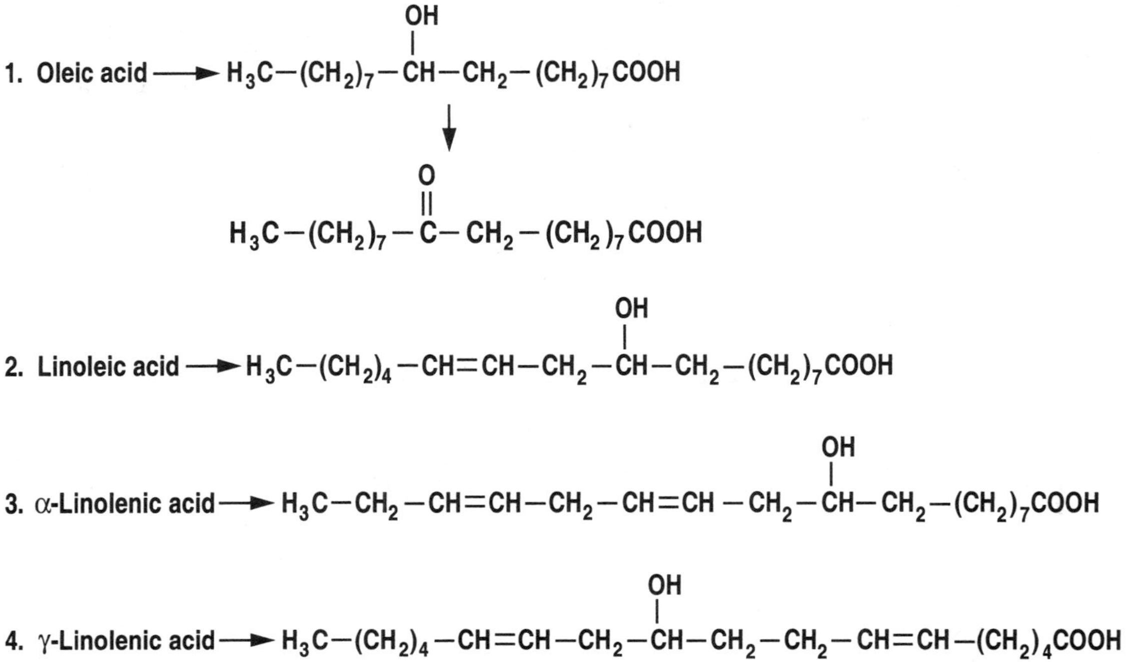

Figure 1. Bioconversion products from unsaturated fatty acids by strain DS5 hydratase.

1. Oleic acid $\longrightarrow$ $H_3C-(CH_2)_7-\overset{OH}{\underset{|}{CH}}-CH_2-(CH_2)_7COOH$ $\longrightarrow$ $H_3C-(CH_2)_7-\overset{O}{\underset{\|}{C}}-CH_2-(CH_2)_7COOH$

2. Linoleic acid $\longrightarrow$ $H_3C-(CH_2)_4-CH=CH-CH_2-\overset{OH}{\underset{|}{CH}}-CH_2-(CH_2)_7COOH$

3. α-Linolenic acid $\longrightarrow$ $H_3C-CH_2-CH=CH-CH_2-CH=CH-CH_2-\overset{OH}{\underset{|}{CH}}-CH_2-(CH_2)_7COOH$

4. γ-Linolenic acid $\longrightarrow$ $H_3C-(CH_2)_4-CH=CH-CH_2-CH-CH_2-CH=CH-(CH_2)_4COOH$

The absolute configuration of DOD was originally determined to be R, R (20) by circular dichroism (CD). Recently, an alternative method to CD was used to determine the absolute configuration of DOD. The method involved formation of the (-)-menthoxycarbonyl (MCO) derivative of the two hydroxyls, oxidative cleavage of the double bond (21,22), and then gas chromatographic analysis of the two methylated diastereomeric acid fragments, methyl-2-MCO-decanoate and dimethyl-2-MCO-octandioate, respectively (Figure 2). As described by previous workers in GC, the 2(S)-MCO derivatives elute at earlier times than the 2(R)-MCO derivatives. Comparing the GC analysis of the MCO derivatives obtained from DOD with that obtained from a partially racemized sample, DOD was determined to be 7(S),10(S)-dihydroxy-8(E)-octadecenoic acid (23).

The production of DOD from oleic acid is unique in that it involves an addition of two hydroxy groups at two positions and a rearrangement of the double bond of the substrate molecule. Subsequent investigation of reactions catalyzed by PR3 led to the isolation of another new compound, 10-hydroxy-8-octadecenoic acid (HOD)(24). From the structural similarity between HOD and DOD, it is likely that HOD is an intermediate in the formation of DOD from oleic acid by strain PR3. Kinetic studies (24) showed that the conversion of HOD to DOD is not the rate-limiting step. The bioconversion pathway for the production of DOD from oleic acid was postulated with HOD as the intermediate and that the unsaturation at position 8 was possibly in *cis*. Recently, we determined that the rearranged double bond of HOD was *trans* form by NMR and FTIR analyses (25). The absolute configuration of the hydroxy group at carbon 10 of HOD also was determined to be in the S from the menthoxycarbonyl (MCO) derivatization of the hydroxy group followed by oxidative cleavage of the double bond and methyl esterification (22). This result coincided with our other findings that the main final product DOD represented 7(S),10(S)-dihydroxy configuration (23). In addition, a minor isomer of HOD (about 3%) with 10(R) configuration was also detected. The overall bioconversion pathway of oleic acid to DOD by strain PR3 is shown in figure 3a. Substrate (oleic acid) is first converted to HOD by PR3 during which one hydroxyl group is introduced at C10(S) and a double bond is shifted from C9 *cis* to C8 *trans*. This suggests that there are at least two or more enzymes involved in this first step for *cis-trans* shifted isomerization of double bond and further hydroxylation introducing a hydroxyl group at C7(S). Hou (26,27) reported that a C10 position-specific and *cis*-specific hydratase was involved in the hydration of unsaturated fatty acid by *Flavobacterium* sp. DS5. It was proposed that the introduction of C10 hydroxyl group with the removal of C9 *cis* double bond was typical for the hydration reaction of unsaturated fatty acids. It is unlikely that a hydratase is involved in the reaction by PR3 in that the C9 double bond of the substrate is retained as a shifted *trans*-configuration during the hydroxylation followed by the formation of DOD.

Strain PR3 also converted ricinoleic acid to a more polar compound in 35% yield (19). The structure of this new product was determined by GC/MS, FTIR, and NMR

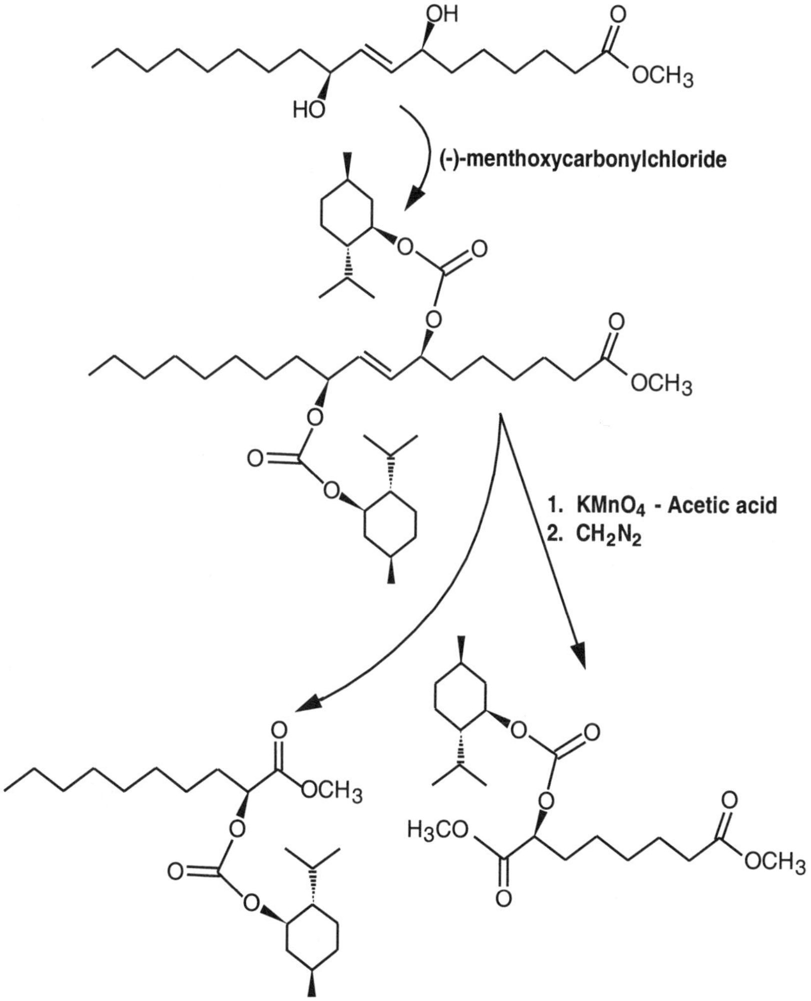

Figure 2. Method of producing (-)-menthoxycarbonyl (MCO) derivatives for chiral analysis by GC.

98

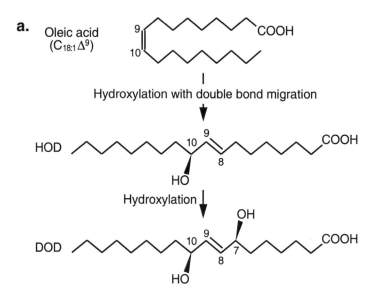

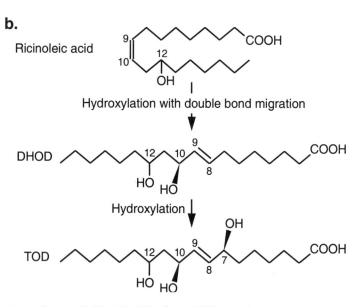

Figure 3. Biosynthesis of (a) 7,10-Dihydroxy-8(E)-octadecenoic acid produced from oleic acid; (b) TOD from ricinoleic acid, by *Pseudomonas aeruginosa* PR3.

to be 7,10,12-trihydroxy-8(E)-octadecenoic acid (TOD). The reaction mechanism is thought to be the same as that for the conversion of oleic acid to DOD. We also isolated as an intermediate, 10,12-dihydroxy-8(E)-octadecenoic acid, in the production of TOD (Figure 3b, unpublished data). Physiological activity tests showed that DOD has activity against the growth of *Bacillus subtilis* and a common pathogen, *Candida albican.*

Mercade *et al.* (28) reported a similar type of compound, dihydroxyoctadecenoic acid produced by *Pseudomonas* 42A2. However, the positions for the double bond and hydroxy groups of their compound were determined in a later report (29,30). Recently, this group reported the oxidation of oleic acid to (E)-10-hydroperoxy-8-octadecenoic acid and (E)-10-hydroxy-8-octadecenoica acid in addition to DOD by 42A2 (30).

III. TRIHYDROXY UNSATURATED FATTY ACID

Recently, we reported the production of a new compound, 12,13,17-trihydroxy-9(Z)-octadecenoic acid from linoleic acid by a new microbial isolate. The microorganism, which performs this unique reaction, was isolated from a dry soil sample collected at McCalla, Alabama. The organism, a Gram (+), nonmotile rod (0.5μm X 2μm), was identified as *Clavibacter* sp. ALA2 (31).

The structure of the new compound was determined by MS, FTIR and NMR. The chemical ionization mass spectrum of the methyl ester prepared with diazomethane gave a molecular ion of m/z 345. Fragments of 327 (M-18), and 309 (M-2 x 18) also were seen. The electron impact spectrum of the methyl ester produced ions corresponding to α-cleavage. Ions at m/z 227 (25%) and 129 (100%) placed two hydroxy groups at the C-12 and C-13 positions and the third hydroxy group at a position higher than C-13. Proton and ^{13}C NMR analyses further confirmed the structure. Resonance signals (ppm) and corresponding molecular assignments (31) located the three hydroxy groups at C12, C13, and C17 and established the identity of the product as 12,13,17-trihydroxy-9(Z)-octadecenoic acid (Figure 4). The coupling constant of 10.7 Hz at C9,10 confirmed our infrared data that the unsaturation had the *cis* configuration (31).

Production of trihydroxy unsaturated fatty acids in nature is rare. Similar compounds reported are all produced in trace amounts by plants. Kato *et al.* (32,33) reported that hydroxy and epoxy unsaturated fatty acids present in some rice cultivars acted as antifungal substances and are active against rice blast fungus. Recently, mixed hydroxy fatty acids were isolated from the *Sasanishiki* variety of rice plant, which suffered from the rice blast disease, and were shown to be active against the fungus (34). Their structures were identified as 9S,12S,13S-trihydroxy-10-octadecenoic acid and 9S,12S,13S-trihydroxy-10,15-octadecadienoic acid (46,47). Similarly, 9,12,13-Trihydroxy-10(E)-octadecenoic acid isolated from *Colocasia antiquorum* inoculated with *Ceratocystis fimbriata*, showed anti-black rot fungal activity (35).

Other than extraction from plant materials, our discovery (31) was the first report on production of trihydroxy unsaturated fatty acids by microbial transformation, and that the structure of THOA resembles those of plant self-defense substances.

The optimum conditions for the bioconversion of linoleic acid to THOA were pH 7.0

and 30 C (36). The maximum production of THOA was found after 5-6 days of incubation. Further incubation did not reduce THOA content in the medium indicating that strain ALA2 does not metabolize THOA.

The biological activity of THOA at 200 ppm concentration was tested against several plant pathogenic fungi (37). The results, expressed in percent growth inhibition, are listed in Table 1. THOA inhibited the growth of *Erysiphe graminis* (Wheat powdery mildew); *Puccinia recondita* (wheat leaf rust); *Phytophthora infestans* (potato late blight); and *Botrytis cinerea* (cucumber botrytis). It appears that the specificity of trihydroxy fatty acids against certain plant pathogenic fungi may depend on the location of the hydroxyl groups on the trihydroxy fatty acid molecule.

Table I. Antifungal Activity of THOA (200 ppm)

Fungus	Disease	% Inhibition
Erisyphe graminis	Wheat powdery mildew	77
Paccinia recondita	Wheat leaf rust	86
Pseudocercosporella herpotrichoides	Wheat foot rot	0
Septoria nodorum	Wheat glume blotch	0
Pyricularia grisea	Rice blast	0
Rhizoctonia solani	Rice sheath blight	0
Phytophthora infestans	Potato late blight	56
Botrytis cinerea	Cucumber botrytis	63

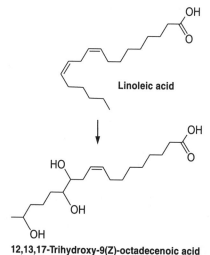

Linoleic acid

12,13,17-Trihydroxy-9(Z)-octadecenoic acid

Figure 4. 12,13,17-Trihydroxy-9(Z)-octadecenoic acid produced from linoleic acid by *Clavibacter sp.* ALA2.

References

1. Naughton, F.C. 1974. *J. Am. Oil Chem. Soc.* 51: 65-71.
2. Wallen, L.L.; Benedict R.G.; and Jackson, R.W. 1962. *Arch. Biochem. Biophys.* 99:249-253.
3. Schroepfer, G. J. Jr. and Block, K. J. 1963. *J. Am. Chem. Soc.* 85: 3310-3315.
4. Schroepfer, G. J. Jr. and Block, K. J. 1965. *J. Biol. Chem.* 240: 54-65.
5. Seo, C. W.; Yamada, Y.;Takada, N. and Okada, H. 1981.*Agric. Biol. Chem.* 45: 2025 –2030.
6. Litchfield, J. H. and Pierce, G. E. U.S. Patent 4,582,804 (1986).
7. Koritala, S.; Hosie, L.; Hou, C. T.; Hesseltine, C. W. and Bagby, M. O. 1989. *Appl. Microbiol. Biotechnol.* 32: 299-304.
8. Davis, E.N.; Wallen, L.L ; Goodin, J.C.; Rohwedder, W.K. and R.A. Rhodes. 1969. *Lipids* 4: 356-362.
9. Kaneshiro, T.; Huang, J-K.; Weisleder, D. and Bagby, M.O. 1994. *J. Ind. Microbiol.* 13:351-355.
10. Blank, W.; Takayanagi, H.; Kido, T.; Meussdoerffer, F; Esaki, N. and Soda, K. 1991. *Agric. Biol. Chem.* 55: 2651-2652.
11. El-Sharkawy, S. H.; Yang, W.; Dostal, L. and Rosazza, J.P.N. 1992. *Appl. Environ. Microbiol.* 58: 2116 - 2122.
12. Yang, W.; Dostal, L. and Rosazza, J. P. N. 1993. *Appl. Environ. Microbiol.* 59: 281-284.
13. Lanser, A. C. 1993. *J. Am. Oil Chem. Soc.* 70: 543-545.
14. Hou, C.T. 1994. *Appl. Environ. Microbiol.* 60: 3760-3763.
15. Hou, C.T. 1994. *J. Am. Oil Chem. Soc.* 71: 975-978.
16. Hou, C.T. and Bagby, M.O. 1991. *J. Ind. Microbiol.* 7: 123-130.
17. Hou, C.T.; Bagby, M.O.; Plattner, R.D. and Koritala, S. 1991. *J. Am. Oil Chem. Soc.* 68: 99-101.
18. Hou, C.T.; Nakamura, L.K.; Weisleder, D.; Peterson, R.E. and Bagby, M.O. 1993. *World J. Microbiol. Biotechnol.* 9: 570-573.
19. Kuo, T.M.; Manthey, L.K. and Hou, C.T. 1998.*J. Am. Oil Chem. Soc.* 75:875-879.
20. Knothe, G.; Bagby, M. O.; Peterson, R. E. and Hou, C. T. 1992. *J. Am. Oil Chem. Soc.* 69: 367-371.
21. Hamberg, M. 1971. *Anal. Biochem.* 43: 515-526.
22. Hamberg, M.; Herman, R.P. and Jacobson, U. 1986. *Biochem. Biophys. Acta.* 879: 410-418.
23. Gardner, H.W. and Hou, C.T. 1999. *J. Am. Oil Chem. Soc.* (in press).
24. Hou, C.T. and Bagby, M.O. 1992. *J. Ind. Microbiol.* 9: 103-107.
25. Kim, H.; Gardner, H.W. and Hou, C.T. 1999. *J. Am.Oil Chem. Soc.* (submitted).
26. Hou, C.T. 1995. *J. Ind. Microbiol.* 14: 31-34.
27. Hou, C.T. 1995. *J. Am. Oil Chem. Soc.* 72: 1265-1270.

28. Mercade, E.; Robert, M.; Espuny, M. J.; Bosch, M. P.; Manreesa, M. A.; Parra, J. L. and Guinea, J. 1988. *J. Am. Oil Chem. Soc.* 65: 1915-1916.

29. de Andres, C.; Mercade, E.; Guinea, J. and Manresa, A. 1994. *World J. Microbiol. Biotechnol.* 10: 106-109.

30. Guerrero, A.; Casals, I.; Busquets, M.; Leon, Y. and Manresa, A. 1997. *Biochim. Biophys. Acta* 1347: 75-81.

31. Hou, C.T. 1996. *J. Am. Oil Chem. Soc.* 73:1359-1362.

32. Kato, T.; Yamaguchi, Y.; Abe, N.; Uyeharaa, T.; Nakai, T.; Yamanaka, S. and Harada, N. 1984. *Chem. Lett.* 25: 409-412.

33. Kato, T.; Yamaguchi, Y.; Uyehara, T.; Yokoyama, T.; Namai, T. and Yamanaka, S. 1983. *Tetrahedron Lett.* 24: 4715-4718.

34. Kato, T.; Yamaguchi, Y.; Abe, N.; Uyehara, T.; Namai, T.; Kodama, M. and Shiobara, Y. 1985. *Tetrahedron Lett.* 26: 2357-2360.

35. Masui, H.; Kondo, T. and Kojima, M. 1989. *Phytochemistry* 28: 2613-2615.

36. Hou, C.T.; Brown, W.; Labeda, D.P.; Abbott, T.P. and Weisleder, D. 1997. *J. Ind. Microbiol. Biotechnol.* 19: 34-38.

37. Hou, C.T. 1998. SIMB, annual meeting, Denver, CO. p2.

Chapter 8

The Nicotinamide Cofactors: Applications in Biotechnology

Hugh O'Neill and Jonathan Woodward

Chemical Technology Division, Oak Ridge National Laboratory[1], Oak Ridge, TN 37831–6194

The applications for $NAD(P)^+$-dependent oxidoreductases are constantly being expanded. For a process to be economically viable, the regeneration and retention of the coenzyme in a continuous reaction system are very important. This article deals with the strategies that have been employed for the regeneration and retention of NAD(P)(H) in such systems. The utilization of $NAD(P)^+$-dependent dehydrogenases in biotechnological processes is also discussed. Although enzymatic cofactor regeneration procedures have traditionally proved most effective the development of alternative methods, such as electroenzymatic techniques and the use of chemical redox mediators, demands consideration.

General Introduction

The function of pyridine coenzymes is related to the unique combination of functional groups that compose these molecules (Fig. 1). The nicotinamide moiety is the oxidation-reduction center. The remaining portion of the molecule is important for selective interactions with enzymes. This is adenosine diphosphoribose in nicotinamide adenine dinucleotide (NAD^+). Nicotinamide adenine dinucleotide phosphate ($NADP^+$) has an additional phosphate group esterified to the 2' hydroxyl of the adenine (adenosine diphosphoribose phosphate). In vivo, the functions of NAD^+ and $NADP^+$ are distinct. NAD^+ is an electron acceptor in the oxidation of fuel molecules. The reduced form (NADH) is oxidized by the respiratory chain to generate adenosine 5'-triphosphate (ATP). In contrast, the reduced form of $NADP^+$ (i.e., NADPH) serves as an electron donor in reductive biosynthesis. Unlike other electron-

[1]Managed by Lockheed Martin Energy Corp. under contract DE–AC05–96OR22464 with the U.S. Department of Energy.

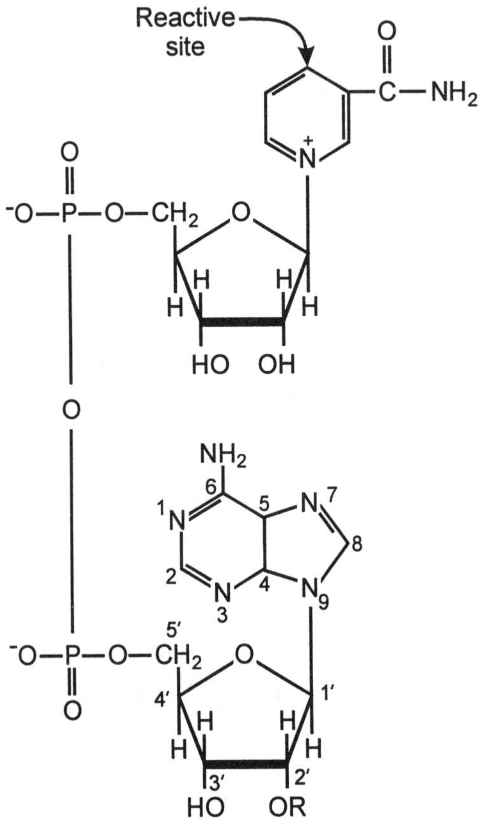

Figure 1. Structure of the oxidized form of nicotinamide adenine dinucleotide (NAD⁺) and of nicotinamide adenine dinucleotide phosphate (NADP⁺). In NAD⁺, R = H; in NADP⁺, R = PO₃²⁺.

Figure 1. Structure of the oxidized form of nicotinamide adenine dinucleotide (NAD^+) and of nicotinamide adenine dinucleotide phosphate ($NADP^+$). In NAD^+, R = H; in $NADP^+$, R = PO_3^{2+}.

transfer centers such as Cu, heme, and flavin, these molecules exist free in solution and function by binding to enzymes transiently during enzymatic oxidation and reduction.

The role of the pyridine nucleotides has become increasingly important to the biotechnology industry. They are used in various analytical, biomedical, and technological processes, where they take part in a wide range of oxidation-reduction reactions via pyridine-dependent dehydrogenases that lead to the synthesis of fine chemicals for the pharmaceutical and food industries. These oxidoreductase enzymes catalyze many of the reactions that are difficult to carry out by conventional chemistry.

Presently, the main issue with the use of coenzymes in preparative chemistry is an economic one. NAD(P)(H) is an unstable and expensive commodity. Stoichiometric amounts of cofactor are consumed in dehydrogenase-catalyzed reactions. Therefore, efficient methods are required for the regeneration of the reduced or oxidized cofactor for process economy (Fig. 2). As an example, the enzymatic conversion of acetophenone to sec-phenethyl alcohol by an $NADP^+$-dependent alcohol dehydrogenase may be considered. The reduction of 1 mol of acetophenone requires 1 mol of NADPH. Table I shows how the overall cost of the process becomes lower as the number of times that NADPH is recycled during the reaction increases.

**Table I. Effect of Recycle Number on the NADPH Costs
for the Production of 1 kg of sec-Phenethyl Alcohol**

Recycle Number	NADPH Costs ($)
1 (no recycle)	43,400
10	4,340
100	434
1,000	43
10,000	4.3
100,000	0.4
250,000	0.16
500,000	0.08

NOTE: These figures are based on a cost of $5000/kg NADPH.
SOURCE: Reproduced from Industrial Enzymology, 2nd edition, eds. T. Godfrey and S.I. West, 1996, p. 159.

Besides the decrease in costs, the regeneration of cofactors can shift the position of equilibrium toward product formation. The need for stoichiometric amounts of the cofactor is eliminated, which can simplify downstream processing. More importantly, such a process eliminates the problem of product inhibition by consuming the cofactor produced in the synthetic reaction as it is formed. An efficient regeneration system will fulfill certain criteria (1,2). The enzymes, reagents, and equipment must be inexpensive and readily available. The total turnover number of the cofactor (TTN =

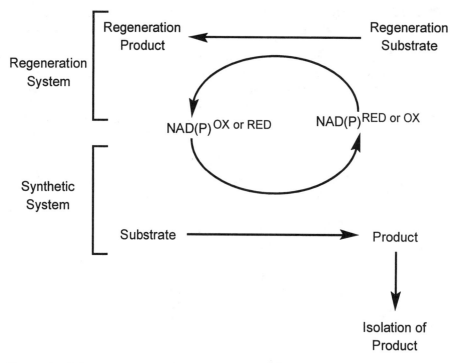

Figure 2. Schematic diagram of an enzymatic synthesis coupled to a cofactor regeneration system.

mol product formed/mol cofactor) should be high. For example, a good regeneration system should exceed a TTN value of 1000 *(3)*. Cofactor regeneration should facilitate the thermodynamically favorable formation of product. The reagents and by-products should not interfere with product isolation and be compatible with the components of the synthetic system. The extent of the transformation should be easily monitored. Regeneration schemes that lead to an enzymatically inactive cofactor should be avoided.

Four primary methods are used for cofactor regeneration: enzymatic, electrochemical, biological, and chemical/photochemical *(1)*. Enzymatic methods are preferred because of their high reaction rates, reaction selectivity, and compatibility with enzymatic synthesis systems. In principle, electrochemical techniques would be the best means of cofactor regeneration because no addition of electron donors or acceptors is needed and no by-products are formed. However, in practice, fundamental difficulties exist with these systems, which are discussed later. Biological and chemical/photochemical methods are not widely used and have been dealt with elsewhere *(1)*. This article concentrates on the enzymatic systems but also discusses some recent advances in electrochemical systems.

The usefulness of a coupled reaction system is dictated primarily by the TTN. A reaction that has a high TTN indicates an efficient turnover of the cofactor with respect to the amount of cofactor initially added to the system and the amount of product generated. The type of reactor that is employed will affect this value. In a batch reactor, the maximum value of TTN is set by the maximum allowed substrate concentration divided by the minimum cofactor concentration required to effect acceptable reaction rates. The factors that influence this value are substrate solubility, stability of the enzyme(s), product inhibition, and rate of inactivation of the cofactor during the reaction compared with its utilization and regeneration. In a continuous-flow reactor, the TTN can be increased if the coenzyme is retained in the reactor while substrate is continually added with the simultaneous removal of product. A large part of this chapter deals with methods that have been employed to retain cofactors in continuous-flow bioreactors.

The stability of coenzymes also demands consideration. The stability of these molecules is affected primarily by pH and temperature *(4,5)*. However, buffer salt composition also plays a role *(6)*. NADH and NADPH are very stable in dilute alkali media. Conversely, the oxidized forms of these molecules are very stable in dilute acid. At intermediate pH values, the reduced forms become unstable as the pH is decreased, and the oxidized forms become unstable as the pH is increased. The rate of decomposition is further accelerated as the temperature is increased. Furthermore, several salts such as phosphate, acetate, NaCl, and KCl have also been shown to destabilize these coenzymes *(6,7)*. The major breakdown products of the decomposition of NADH are ADP ribose and nicotinamide, which are inhibitors of some dehydrogenases *(8,9)*. The pH for optimal stability of a steady-state mixture of $NAD(P)^+$ and NAD(P)H was estimated *(1)*. In an unbuffered solution, the minimum destruction of a 9:1 steady-state mixture of NAD^+ and NADH was calculated to be pH 7.0. In HEPES buffer, this was increased to pH 7.5. For a 1:1 ratio in unbuffered or in HEPES solution, the destruction was minimal at pH levels of 7.7 and 8.5, respectively. The stability of NAD(P)H is less dependent on the buffer composition than NAD(H). In both unbuffered and buffered solutions, the pH for the minimal destruction of either cofactor mixture was calculated to be pH 8.5. In addition, the oxidized cofactors are also susceptible to nucleophilic attack at the pyridinium ring by many compounds, including phosphate and pyruvate, which leads to the inactivation of the cofactor *(1)*.

Enzymatic Methods for the Regeneration of Cofactors

The enzymatic regeneration of cofactors has been investigated extensively. Many methods have been used *(1,10)* with varying degrees of success. The most popular systems are presented here.

Regeneration of NAD(P)H

The formate/formate dehydrogenase (FDH) system for NADH regeneration has been used in the preparative synthesis of L-amino acids and D-, L-α-hydroxy acids *(11)* (Fig. 3A). Preparations of FDH from *Candida boidinii (12)* and *Pseudomonas oxalaticus* are commercially available. The enzyme is stable between pH 6.5 and 7.5 if it is protected from autooxidation. Formate and its metabolic product CO_2 are nontoxic to enzymes and NAD(H). Formate is a very good reducing agent and drives reduction of NAD^+ to completion. The removal of the CO_2 produced makes the reaction irreversible. The main disadvantages of this regenerative system are the initial expense of the enzyme; its low specific activity, which means that a relatively high concentration of enzyme is required; and the fact that it does not utilize $NADP^+$.

Glucose 6-phosphate dehydrogenase (G6PDH) from *Leuconostoc mesenteroides* catalyzes the reduction of $NADP^+$ and NAD^+ (Fig. 3B). This enzyme has been used in the batch synthesis of chiral α-hydroxy acids and alcohols *(13,14)*. The substrate glucose 6-phosphate and the product 6-phosphogluconolactone are not toxic to enzymes. The reduction of $NAD(P)^+$ is strongly favored due to the spontaneous hydrolysis of the product to 6-phosphogluconate. Although the enzyme is available commercially and is inexpensive, glucose 6-phosphate is an expensive reagent *(1)*. Both glucose 6-phosphate and 6-phosphogluconate catalyze the hydration of the nicotinamide of NAD(P)H, causing inactivation of the coenzyme *(13)*. Glucose 6-sulfate has been proposed as an alternative to glucose 6-phosphate. However, only yeast G6PDH shows acceptable rates with this substrate.

Glucose dehydrogenase (GDH) from *Bacillus cereus* exhibits high specific activities for NAD^+ and $NADP^+$ *(15)* (Fig. 3B). Its substrate, glucose, is a strong reducing agent and does not catalyze the hydration of the reduced cofactor. In addition, the product glucono-δ-lactone is spontaneously hydrolyzed to gluconate, which makes the reaction irreversible. Immobilization of the enzyme also increases its thermal stability *(16)*. A possible disadvantage of this system is the separation of the desired product from gluconate at the end of the reaction. Also, this enzyme is not catalytically active with polyethlene glycol (PEG)–NAD(P)$^+$ derivatives *(17)*.

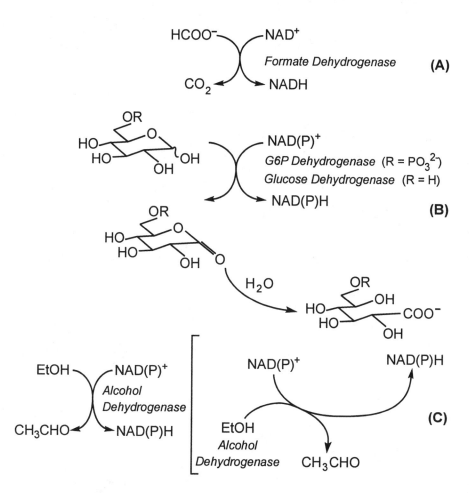

Figure 3. Enzymatic reactions for the regeneration of NAD(P)H.

Various methods have been proposed for the regeneration of NADH based on the oxidation of alcohols (e.g., ethanol and isopropanol) by alcohol dehydrogenase (Fig. 3C). The inexpensive starting material and the volatility of ethanol and its product acetaldehyde make this system attractive. However, some disadvantages prevent its general use. Both ethanol and acetaldehyde deactivate enzymes, and a large excess of ethanol is required because it is only weakly reducing *(1)*. In addition, acetaldehyde, even at low concentrations, inhibits alcohol dehydrogenase from horse liver and yeast. It also condenses with NADH to give 1,2-dihydro-2-ethylene nicotinamide *(1)*. For this regeneration system to be useful, it is necessary to remove the acetaldehyde as it is produced.

Regeneration of NAD(P)$^+$

The α-ketoglutarate/glutamate dehydrogenase (α-KG/GluDH) system is the most useful one for the regeneration of oxidized cofactor *(10)* (Fig. 4A). Glutamate dehydrogenase (bovine liver) catalyzes the oxidation of NAD(P)H under anaerobic conditions. The reagents and enzymes are both stable and innocuous, and the potential for the reductive amination of α-KG is favorable for most biochemical oxidations. The disadvantages of this system are the moderate specific activity of glutamate dehydrogenase and the fact that the glutamate formed must be separated from the desired product.

The pyruvate/lactate dehydrogenase (LDH) (rabbit muscle) system offers some advantages over the α-KG/GluDH system in terms of cost of materials, enzyme stability, and higher specific activity (Fig. 4B). However, the pyruvate condenses with itself in aqueous solution and reacts with NAD$^+$ by nucleophilic substitution *(18)*. Furthermore, this reaction is catalyzed by LDH *(1)*. Glyoxylate, which is reduced to 1-hydroxyacetate by LDH *(1)*, offers some advantages over pyruvate as an oxidizing agent because: (1) it is less expensive, (2) it does not react with NAD$^+$, and (3) the equilibrium constant for its reaction with LDH is higher than that for the reductive amination of α-KG by GluDH. However, LDH recognizes glycolate, the hydrated form of glyoxylate, and catalyzes its oxidation to oxalate in the presence of NAD$^+$. With careful control of the reaction conditions, this side reaction could probably be minimized *(2)*.

The advantage of the acetaldehyde/alcohol dehydrogenase system is the low cost of acetaldehyde and alcohol dehydrogenase (yeast) and the high specific activity of the enzyme (Fig. 4C). However, the potential exists for the deactivation of enzymes by acetaldehyde and ethanol. Also, acetaldehyde is unstable in solution and can both self-condense and condense with NAD$^+$. One possibility is to maintain the acetaldehyde at low concentrations by its continuous addition *(18)*.

Macromolecular Derivatives of Cofactors

In general terms, there are two reasons to immobilize cofactors, and the type of matrix used will dictate the application of the immobilized coenzyme. Solid supports are most commonly used in applications for which the binding properties of the

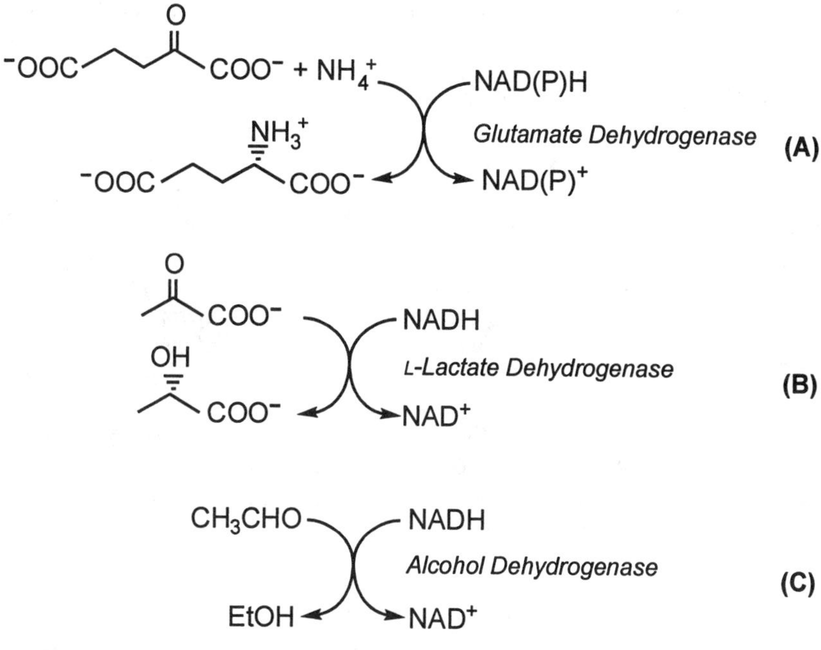

Figure 4. Enzymatic reactions for the regeneration of NAD(P)⁺.

immobilized coenzyme or its analogue are exploited, for example, affinity chromatography. However, if an active immobilized cofactor is required, soluble supports are favored. This chapter deals solely with the techniques that are used to synthesize immobilized coenzymes that retain catalytic activity.

Given the chemical complexity of the $NAD(P)^+$ molecule, many different functional groups are available for chemical modification. To date, all the major portions of the molecules have been modified in an effort to determine their mechanisms of action (19). Many of these derivatives have been observed to function as coenzyme. in dehydrogenase-catalyzed oxidation-reduction reactions. Their function is much less affected by alterations in the purine moiety than by the pyridinium moiety (19). For example, small changes in the nicotinamide mononucleotide moiety have drastic effects on coenzymic activity while radical alterations to the adenosine-5'-monophosphate moiety can occur without total loss of coenzyme function. Crystallographic information is available for several dehydrogenases, which shows that the adenine ring lies in a hydrophobic pocket and is exposed to the solvent (20). Therefore, modification of this portion of the molecule is the most popular technique for immobilization procedures. The substitutions around the AMP portion give rise to variations in the K_m and V_{max} values because the orientation of the purine part of the molecule is important for the retention of full catalytic activity (2). Modifications will invariably interfere with one or both of these parameters. Two positions on the adenosine portion of the molecule are possibly available for chemical modification: the N^6 and the C8 positions (see Fig. 1).

Two approaches are commonly used to generate macromolecular coenzyme derivatives. The first approach involves the introduction of a functional group into the coenzyme molecule followed by a cross-linking reaction with an activated matrix. This has been described as the modular or preassembly approach (21). The advantage of this method is that a chemically defined, homogeneous population of ligands is obtained and no ambiguity exists concerning the group through which coenzyme attachment occurs. However, these types of procedures usually require laborious multistep chemical syntheses and the yields of product are generally quite low. In the second approach, the modified coenzyme is directly attached to a matrix that has a suitable functional group. The major disadvantage of this method is that a heterogeneous population of chemically undefined ligands may be obtained. Thus, depending on the site of attachment, not all of the coenzyme molecules will retain activity. Most of the methods that are described were developed for the immobilization of NAD^+. However, many of these procedures are also applicable to $NADP^+$.

Compared with the native coenzyme, the immobilized cofactor must retain its coenzyme activity. Therefore, it is important that the immobilized coenzyme be available for interaction with an enzyme active site without steric hindrance due to unfavorable properties of the matrix—such as charge, hydrophobicity, hydrophilicity, or size—that may restrict access. Also, the linkage between the coenzyme and the polymer must be stable under the reaction conditions employed.

Modification of the N^6 Position of the Adenine Ring

The N^6 position of the adenine ring is a popular choice for the derivatization of NAD(P)$^+$. The attachment of a reactive functional group at this position has been used as a first step in many immobilization procedures. The classic strategy for the synthesis of N^6-functionalized NAD(P)$^+$ has four steps *(22)*. The first step is alkylation of the N(1) position of the adenine ring to introduce a reactive group leading to N(1)-functionalized NAD(P)$^+$. The N(1)-alkylated derivative is intrinsically unstable. Therefore, the second step involves chemical reduction of NAD(P)$^+$ with sodium dithionite leading to N(1)-functionalized NAD(P)H. This is followed by a Dimroth rearrangement of the N(1)-functionalized NAD(P)H to N^6-functionalized NAD(P)H under harsh alkaline conditions to obtain a more active and chemically stable coenzyme derivative. The last step of the process is the reoxidation of the NAD(P)H derivative either by chemical or enzymatic means. This is necessary to allow further purification to be carried out in acidic media. The by-products of the reaction that do not possess the correct configuration are subsequently degraded in the resulting acidic media and so are separated from the desired product *(23)*.

N^6-carboxymethyl NAD$^+$ and NADP$^+$ have been synthesized using iodoacetic acid *(24)*, 3-propiolactone *(25,26)*, and 3-iodopropionic acid *(27)* as alkylating reagents. The alkylating reagent 3,4-epoxy butanoic acid has been used to synthesize N^6-2-hydroxy-3-carboxypropyl-NAD$^+$ and NADP$^+$ *(28,29)*. Ethyleneimine was used to synthesize N^6-aminoethyl NAD$^+$ *(30)*. The synthesis of an aziridine N^6-2-aminoethyl-NAD$^+$ has also been reported *(31–33)*. Diepoxy compounds such as 1,2,7,8-diepoxy octane are used to introduce epoxy functional groups *(32)*.

The main disadvantage of this method is that the yields are typically low (10–30%) due to losses during the purification procedure after each step. The highest recorded yield for the classical procedure is 50% for the synthesis of N^6-(2-carboxyethyl)-NAD$^+$ *(26)*. A simpler procedure has been devised for the synthesis of N^6-aminoethyl NAD$^+$, which involves two chemical steps and a purification procedure *(22)*. The C(6)-N-alkyl bonds are very stable under the normal operating conditions for enzymatic reactions. These derivatives have short spacer arms that are terminated with an amino or a carboxyl functional group. Diaminohexane and diaminopropane are suitable reagents for extending the length of the spacer arm of N^6-carboxyalkyl-NAD(P)$^+$ derivatives in a carbodiimide-promoted reaction *(34)*.

Modification of the C8 Position

The carbon atom at position 8 of the purine nucleus has the highest electron density and is susceptible to direct electrophilic substitution. It is therefore easier to synthesize C8-substituted analogues than N^6-substituted analogues. Direct bromination of the nucleotide at position 8, followed by nucleophilic displacement with a compound such as diaminohexane, has been achieved with NAD$^+$ and NADP$^+$ *(36,37)*. The reaction of 8-Br-NAD$^+$ with 3-mercaptopropionic acid resulted in the production of 8-(2-carboxyethyl)-thio-NAD$^+$ *(37)*. However, N^6-modified NAD(P)$^+$ derivatives usually interact better with dehydrogenases than do C(8)-modified NAD$^+$

derivatives, b .cause structurally more favorable access to the adenine binding site is maintained (2).

Preparation of Macromolecular Coenzymes.

The preassembly strategy (21) for preparation of macromolecular NAD(P)(H) involves the cross-linking of the functionalized coenzyme dervivatives with soluble polymers to obtain a chemically defined homogeneous population of cross-linked coenzyme. Some polymers such as polyethyleneimine (PEI) (38) and poly-L-lysine (39) have reactive functional groups and can be directly cross-linked with coenzyme derivatives. The most popular polymers for cross-linking coenzyme derivatives, dextran and PEG, are essentially uncharged and unreactive. It is necessary to introduce functional groups into these polymers before cross-linking [e.g. glycylglycylglycyl-dextran(40), cyanuric chloride–activated dextran(41), carboxymethyl-dextran (42), amino-PEG (43), and carboxy-PEG (42,44)]. It is of interest to note that coenzyme derivatives that are prepared using cyanogen bromide are prone to leakage because of the instability of the isourea linkage between the coenzyme derivative and dextran (45). Casein (46), carboxypolyvinylpyrrolidone (44), and maleic anhydride–containing polymers (44) have also been used as support materials.

1-ethyl-(3-dimethylamino-propyl) carbodiimide HCl (EDC), a water-soluble carbodiimide is a suitable reagent for cross-linking a molecule with a carboxyl functional group to one that has an amino functional group. The reaction is carried out in aqueous solution at pH 4.5–4.8 and results in the formation of an amide bond between the reactive species. It was found that cross-linking an NH_2-functionalized coenzyme with an activated carboxylated polymer is much more efficient than reacting an activated COOH-functionalized coenzyme with a polymer possessing an NH_2 functional group (29,42). In the latter reaction, a molecular excess of the coenzyme derivative is necessary. Therefore, further purification steps are required to separate the unreacted polymer from the cross-linked product (43). Carbodiimide coupling reagents are not suitable for use with $NADP^+$ derivatives that have either an amino or a carboxyl functional group because 2', 3'-cyclization of the 2' phosphate occurs at the ribose (2,47).

N-hydroxy-succinimide can also be used as a cross-linking reagent for coenzyme derivatives that have a free primary amine group (48). The reaction is carried out at pH 7.0–8.0 in a nonaqueous environment. It is therefore not suitable for use with carboxylated coenzyme derivatives as they are not soluble under these conditions. However, it can be used for cross-linking $NADP^+$ derivatives with free amino groups.

Direct Methods

The previously described methods require that the coenzyme be modified to introduce a reactive functional group prior to cross-linking to a polymer. However, several methods bypass the need for this step, and the unmodified coenzyme is attached directly to the polymer. The yield of cross-linked polymer is potentially higher in such a procedure because no multistep synthesis is involved. However, a major disadvantage of this type of method is that a heterogeneous population of ligands may be obtained.

The direct attachment of NADH onto a water-soluble acrylic polymer that contained epoxy functional groups has been reported *(49)*. This involved alkylation of the NAD^+ at $N(1)$, reduction of the nicotinamide with sodium dithionite, and Dimroth rearrangement of the alkyl linkage from $N(1)$ to the N^6 position. A similar copolymer that also contained epoxy functional groups was used to cross-link NADH in a single-step reaction. The densities of NADH on the two polymers were 21% (w/w) and 3% (w/w), respectively. However, only 60% of the immobilized cofactor could be oxidized. This indicates that not all of the immobilized cofactor interacts with enzyme, perhaps due to unknown side reactions.

A macromolecular NAD^+ derivative was also prepared by cross-linking NAD^+ to the carboxyl group of alginic acid using EDC *(50,51)*. Different densities of cross-linked NAD^+ were obtained depending on the amount of NAD^+ starting material that was used [3.4–17.6% (w/w)]. The immobilized coenzyme retained approximately 85% of its activity compared with the free coenzyme in an alcohol dehydrogenase–based assay.

Catalytic Activity of Macromolecular Coenzymes

As outlined previously, numerous methods are now available for the immobilization of cofactors. The assessment of the effectiveness of a cofactor is based on its ability to take part in an enzymatic reaction as expressed in terms of the kinetic parameters V_{max} and K_m.

The activity of macromolecular coenzyme derivatives is highly dependent on the enzyme that is used for measurement. For instance, GDH from *Bacillus megaterium* and mannitol dehydrogenase from *Sacchromyces cerevisae* have very little activity with PEG derivatives of $NAD(P)^+$ *(17,43)*. In contrast, the catalytic efficiency of formate dehydrogenase with PEG derivatives is higher than that with the native coenzyme. The same is true of GDH from *B. megaterium* with PEI-NH-succinyl NAD^+ *(17)*.

Buckmann and Carrea *(2)* compiled the enzymatic data for derivatives of NAD^+ that had been synthesized by the preassembly approach and linked at the N^6 position. The most popular enzyme derivatives used to determine coenzyme activity are alcohol dehydrogenase (yeast and horse liver), LDH, malate dehydrogenase, and glutamate dehydrogenase. It was concluded that, in general, the immobilization of cofactors resulted in a decrease in V_{max} and a corresponding increase in K_m.

The factors that affect these parameters are the position of functionalization of the coenzyme, the properties of the spacer group that joins the coenzyme and the matrix, and the nature of the matrix *(2)*. The molecular weight and the density of the immobilized cofactor are also important *(2)*. NAD^+-dependent dehydrogenases are most active with N^6 derivatives. C(8) derivatives are generally less active, and $N(1)$ derivatives are poor substrates for NAD^+ dehydrogenases. The opposite is true of the $NADP^+$-dependent enzymes, which show high activity with the $N(1)$ derivatives.

PEG-linked derivatives are reported to give the best results with most enzymes, while PEI gives erratic results. For example, PEI is highly active with LDH and only weakly active with alanine dehydrogenase. Dextran-NAD$^+$ derivatives retain 8.5–37% of the activity reported for the native coenzyme *(32)*. The other properties of the coenzyme derivatives such as the properties of the spacer arm, the size of the polymer, and the density of the NAD$^+$ are highly dependent on the enzyme used for the investigation.

The kinetic parameters of the coenzymes that are synthesized by simplified strategies are similar to those synthesized by the preassembly approach. Fuller et al. *(49)* reported that the V_{max} and the K_m values of the two NAD$^+$ water-soluble acrylic polymers were, respectively, one-third and twice those of the free coenzyme using five different dehydrogenases. In the case of the NAD$^+$–alginic acid polymer, the reported activity is 85% of the free coenzyme in alcohol dehydrogenase and glucose 6-phosphate dehydrogenase *(50)*.

Water Soluble Enzyme-Coenzyme Conjugates

An alternative approach to coenzyme immobilization is to directly attach the coenzyme to the enzyme. This, in effect, converts the coenzyme to a covalently bound prosthetic group. For a useful conjugate to be obtained, it is important that the coenzyme interact with the active site of the enzyme and also be available for interaction with an unmodified enzyme or redox compound to facilitate its regeneration. To date, only the construction of NAD$^+$ enzyme conjugates has been described.

Persson et al. *(52)* described a novel empirical method for the construction of a GDH-NAD$^+$ conjugate. A functional group was introduced on the surface of the protein in the vicinity of the active site that facilitated the attachment of a coenzyme derivative. Cysteine is considered an attractive amino acid residue for this purpose, because it is found infrequently on the surface of proteins *(53)*. Cysteine is especially useful in this case because no other naturally occurring cysteines are found in GDH from *Bacillus subtilis*. The site-directed mutagenesis technique was used to replace an aspartic acid residue (No. 44) with a cysteine residue. An NAD$^+$ analogue that had a reactive thiol group was made by reacting N^6-[6-aminohexyl-(carbamoylmethyl)]-NAD$^+$ with N-succinimidyl 3-(2-pyridyldithio) propionate (SPDP). This was directly coupled to the Cys^{44} glucose dehydrogenase. The coupling procedure yielded a preparation with one reactive coenzyme molecule per enzyme subunit. The covalent linkage did not cause any loss in the activity of the enzyme or coenzyme. The bound cofactors could all be reduced in the presence of D-glucose, and the specific activity was the same as that of the native enzyme using exogenous NAD$^+$.

The oxidation of glucose by GDHcys^{44}NAD was measured in the presence of a coupled redox system. The coenzyme was recycled in the presence of phenazine methosulfate and 2,6-dichlorophenol-indophenol. The rate constants were independent of the concentration of the complex and showed first-order kinetics, indicating that the reaction between the bound NAD$^+$ and the enzyme is intramolecular. In contrast, the reaction of the enzyme conjugate or native enzyme with free NAD$^+$ in a fixed ratio showed second-order kinetics, indicative of an intermolecular reaction.

The activity of the free coenzyme derivative with the native enzyme was only 5% of that recorded with the native coenzyme. However, the covalently bound coenzyme exhibited a tenfold higher rate than an identical system with native biocatalysts. This is attributed to the high local concentration of coenzyme that is established on immobilization of the coenzymes. It makes the practical application more economical. Furthermore, the stability of the NAD(H) may be increased due to its positioning on the enzyme.

The activity of the GDHcys^{44}NAD conjugate was also examined in a hollow-bed-fiber enzyme bioreactor. Pyruvate and LDH were added for the regeneration of the NAD$^+$. In a continuous flow system a steady-state conversion of pyruvate to lactate of 84% was obtained on the first day. The NAD(H) was recycled 2700 × h^{-1}. A decline in lactate production was seen after the first day. In 2.5 days, a TTN of 135,000 was obtained. The reason for the decline is thought to be the instability of the disulfide bond that linked the enzyme and coenzyme or the inactivation of the coenzyme itself. An NAD$^+$ derivative that has a malamide functional group has also been formed. When this derivative reacts with the sulphydryl groups on the enzyme, it forms an irreversible thioether bond. Preliminary tests in a similar reactor indicate a more stable operation without loss of coenzyme activity.

The construction of similar dehydrogenase-coenzyme complexes was also described. These methods took advantage of naturally occurring εNH$_2$ groups and carboxyl groups on protein surfaces as sites for the attachment of NAD$^+$ derivatives. The characteristics of several of these conjugates are presented in Table II. The data for GDHcys^{44}NAD enzyme are also included for comparison. Their properties have been discussed in detail elsewhere (2). The main disadvantage of this method is that a heterogeneous population of conjugates is obtained because the coupling procedure is random and uncontrollable. This is reflected in the catalytic activities (28–127%) of the conjugates and the amount of bound NAD$^+$ that can be reduced at the active sites (25–90%). The catalytic integrity of the protein part of the conjugates was tested by measuring their relative activities in the presence of exogenous NAD$^+$ compared with the native enzyme. A strategy to direct the attachment of the coenzyme derivative to the active site was conceived whereby a ternary complex was formed between the enzyme, coenzyme derivative, and a pseudo-substrate before the addition of coupling agent. This method proved successful in some cases (58,62,63).

A further extension of enzyme-coenzyme conjugation is the incorporation of an additional catalytic group for the regeneration of the reduced coenzyme. A 5'-ethylphenazine GDH-NAD$^+$ conjugate (EP$^+$-GDH-NAD$^+$) was prepared by linking PEG-5'-ethylphenazine and PEG-NAD$^+$ to GDH (64). This conjugate is a semisynthetic GDH that has glucose oxidase activity. A 5'-ethylphenazine-LDH-NAD$^+$ conjugate (EP$^+$-LDH-NAD$^+$) that has lactate oxidase activity was also constructed (65). The 5'-ethylphenazine moiety acts as a catalytic group for the oxidation of NADH with O$_2$ or MTT [3-(4,5-dimethyl-2-thiazolyl)-2,5-diphenyl-2H-tetrazolium bromide] as electron acceptors. The characteristics and the kinetic properties of these semisynthetic oxidases are presented in Table III.

Table II. Characteristics of Dehydrogenase-NAD$^+$ Conjugates

Enzyme	NAD$^+$ Derivative	Mol Coenzyme / Mol Enzyme Subunit	Substrate Reducible NAD$^+$ (%)	Enzyme Activity % of Native Enzyme	Ref. Number
Liver alcohol dehydrogenase	N^6-[N-(6-aminohexyl)-carbamoylmethyl]-NAD$^+$	0.9–2.1	25	40	54, 55, 57
	N^6-[N-(8-amino-3,6-dioxaoctyl)-carbamoylmethyl]-NAD$^+$	0.4–0.5		52	62
Lactate dehydrogenase	N^6-[N-(6-aminohexyl)-carbamoylmethyl]-NAD$^+$	1.6		28–58	56
	NAD$^+$ activated with N^6-attached diaziumaryl or imidiester groups	0.4–1.1	40–60	80–90	58
Malate dehydrogenase	PEG (M$_r$ 3,000)-N^6-(2-carboxyethyl)-NAD$^+$	0.5–1.2	60–90	77–127	59
Glucose dehydrogenase	PEG (M$_r$ 3,000)-N^6-(2-carboxyethyl)-NAD$^+$	2.1	70	92	60
Formate dehydrogenase	N^6-[N-(6-aminohexyl)-(carbamoylmethyl]-NAD$^+$	0.2 [a]	–	116	61
Cys^{44}glucose dehydrogenase	N^6-[N-(6-aminohexyl)-(carbamoylmethyl]-NAD$^+$	4	100	100	62

NOTE: [a] Calculated taking into consideration only the fraction of bound enzyme reducible by substrate
SOURCE: Adapted from Reference 2.

Table III. Kinetic Properties of Semisynthetic Oxidases

	Mol of Moiety/ Mol Subunit		V$_{max}$ (μM min^{-1})		K$_m$ (mM)		Turnover No. (min^{-1})	
	EP$^+$	NAD$^+$	O$_2$	MTT	O$_2$	MTT	O$_2$	MTT
EP$^+$–GDH–NAD$^+$	0.80	1.20	8.23	2.50	1.57	0.072	24	6.6
EP$^+$–LDH–NAD$^+$	0.46	0.32	3.51	0.42	1.91	0.076	2.3	0.25

The oxidase activity of the EP^+-GDH-NAD^+ is much greater than that of the EP^+-LDH-NAD^+ conjugate. The turnover numbers of the EP^+-LDH-NAD^+ for O_2 and MTT are only 10 and 38%, respectively, of the value recorded for EP^+-GDH-NAD^+ (Table III). This was explained in terms of the higher activity and affinity of PEG-NAD^+ for LDH. The K_m of GDH for PEG-NAD^+ is >4.2 mM, while the K_d values of LDH for PEG-NAD(H) are 0.5 mM and 2.8 µM for the oxidized and reduced forms, respectively. The much higher affinity of NADH for the active site of LDH means that almost all of the NAD^+ moieties are in the reduced form, and only 2.1% of the NADH is in a free state. The rest remains in the active site and is unavailable for interaction with the EP^+. Referred to as the "hide-and-seek effect," this phenomenon greatly reduces the oxidase activity of the EP^+-LDH-NAD^+ by lowering the effective concentration of free NADH. Therefore, two additional catalytic forms of EP^+-LDH-NAD^+ exist that are not in the catalytic cycle of EP^+-GDH-NAD^+ (Fig. 5).

The lower oxidase activity of both conjugates with MTT compared with O_2 is not fully understood but is attributed to steric hindrance between it and the bound EPH (Table III). These conjugates portray three important rate-acceleration mechanisms, the high effective concentration of the substrate, the intramolecular coupling of successive catalytic sites, and the multiple connections between two types of catalytic sites *(65)*.

Electrochemical Methods

Electrochemical methods have been described as alternatives to methods based on enzyme technology. However, the electrochemical reactions of the oxidized and reduced cofactors are kinetically unfavorable at the bare electrode surface *(66)*. High overpotentials are required due to the slow rate of reaction. The direct oxidation of NAD(P)H at conventional electrodes such as C, Pt, or Au has been demonstrated with high yields of active cofactor, typically 90–99.3% *(51,67,68)*. The reduction of NAD(P)$^+$ involves the stereospecific transfer of a hydride moiety. The one-electron reduction of the NAD(P)$^+$ results in the formation of a cation radical that is unstable and forms an enzymatically inactive 4-4' dimer *(69,70)*. Nonselective reduction also results in the formation of 1,2- and 1,6-dihydronicotinamide species. The immobilization of the cofactor on a polymer can suppress this intermolecular radical formation *(51)*.

To carry out efficient electrochemical regeneration of pyridine nucleotides, it is necessary to use a catalytic system that can operate at lower overpotentials compared with the redox potential of the NAD(P)$^+$/NAD(P)H couple. Various redox mediators have been used to catalyze the oxidation of NAD(P)H, including quinones *(71,72)*, phenoxazines that contain quinoidic structures *(73–75)* and phenazines *(76,77)*. The biocatalytic reduction of NAD(P)$^+$ has been achieved by using redox dyes *(78)* and rhodium(III) bipyridyl complexes *(79,80)*. These facilitate hydride transfer to the C(4) position of the nicotinamide moiety. A major problem that limits the use of these systems is electrode fouling by adsorbed components of the solution, including the cofactor.

(A)

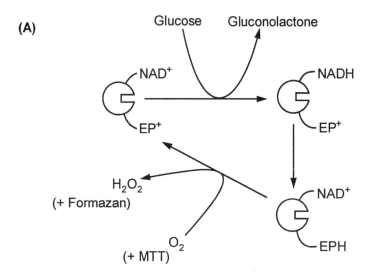

(B)

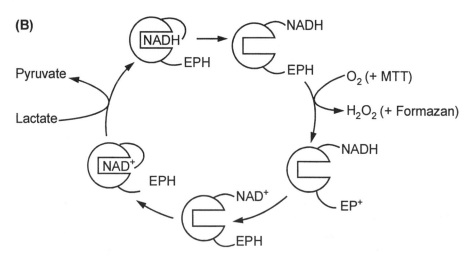

Figure 5. Schematic representations of the catalytic cycles of semisynthetic glucose oxidase (A) and semisynthetic lactate oxidase (B). (Reproduced with permission from references 64 and 65. Copyright 1991/2 FEBS.)

A novel method was reported for the construction of an organized assembly of a catalyst, cofactor, and enzyme on the surface of an electrode *(81)*. The method is based on the assembly of an electrocatalyst, pyroquinoline quinone (PQQ), and an $NAD(P)^+$ monolayer on the surface of a cystamine-functionalized gold electrode. This acts as an active interface for the electrooxidation of reduced dihydronicotinamide cofactor. A cofactor-dependent dehydrogenase is assembled on the NAD^+ unit by associative affinity interactions. This temporary stability of the interactions allows the cross-linking of the enzyme layers using glutaraldehyde to produce a stable, electrically contacted enzyme electrode (Fig. 6).

The properties of an LDH electrode that was constructed in this way were investigated. Calcium ions are required as cocatalysts for the oxidation of NADH by PQQ. It was estimated that the surface coverage of the enzyme on the electrode was 3.5 pmol cm^{-2}. The cross-linked enzyme exhibited a high amperometric stability. No decline in current due to the dissociation of the enzyme from the electrode was noted, nor were the biocatalytic features affected by the cross-linking procedure. The anodic current response was dependent on the concentration of lactate up to a concentration of 10 mM, after which it leveled off, indicating that the active sites of the enzyme were saturated. A K_m of 3.6 mM was calculated for lactate and a maximal rate was I_{max} = 4.1 mA. The addition of exogenous NAD^+ increased the current response twofold, presumably because LDH possesses four NAD^+ binding sites, only two of which are used for the immobilization (Fig. 6). The additional NADH formed diffuses to the electrode, where it is oxidized by PQQ at the electrode surface. The same approach was used to construct an electrode that utilized alcohol dehydrogenase. The anodic response of this electrode was dependent on the concentration of ethanol up to at least 5 mM *(81)*.

Retention of Cofactors in Bioreactors

Ultrafiltration bioreactors are very useful for enzyme catalysis in homogeneous solutions. A continuous supply of substrate can be added with the concomitant removal of products. As mentioned previously with enzyme-coenzyme systems, the TTN can be substantially increased in a continuous-reaction system. Most of the macromolecular derivatives of coenzymes that have been described are retained by commercially available ultrafiltration membranes [molecular weight cut off (MWCO): 5–10 kDa]. This is especially true of bulky polymer-bound coenzyme derivatives, for example, PEI *(2)*. With nonbulky macromolecular coenzyme derivatives such as PEG derivatives, it was shown that PEG 20-kDa derivatives were retained to a higher degree than PEG 10-kDa derivatives in an ultrafiltration apparatus with a MWCO of 5 kDa *(82)*. Losses of 4.3 and 1.7% per day were recorded for PEG 10-kDa and PEG 20-kDa derivatives, respectively.

The retention of native coenzyme in continuous-reaction systems has also been described. Nanofiltration membranes (MWCO: 0.5–0.7 kDa) have been used in the synthesis of xylitol and mannitol *(3)*. However, the flow characteristics of these membranes are very poor. The possibility of substrate or product retention is also a factor to be considered. NAD(P)(H) carries a net negative charge at pH values

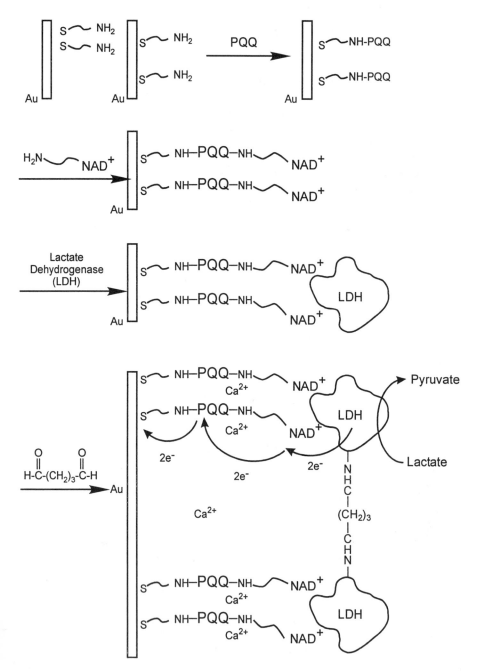

Figure 6. Strategy for the assembly of an integrated electrode. (Reproduced with permission from reference 81. Copyright 1997 American Chemical Society.)

above 3. Negatively charged membranes (MWCO: 1 kDa) have been used to exploit this property of the coenzyme *(83)*. Although the flow characteristics of these membranes are better, negatively charged proteins should be used to prevent membrane fouling. The retention of charged components of the solution also poses problems.

A novel method for the retention of $NAD(P)^+$ in a conventional uncharged 10-kDa ultrafiltration membrane reactor is the addition of PEI, a cationic polymer, to the reaction vessel *(84)*. It interacts electrostatically with the cofactor to keep it in the bioreactor. This method is more suited to the use of $NADP^+$ than NAD^+ because of the additional negative charge on the molecule. The ionic strength of the solution should also be kept low to optimize the interactions. A bioreactor that utilized GDH and LDH for the production of gluconate and lactate achieved a 90% conversion of substrate to product. Coenzyme retention of 80% was reported and the TTN of NAD^+ in a 50-h experiment was 14,190. An additional advantage of this method is that the PEI also confers additional stability on the GDH and LDH *(85)*.

Biotechnological Applications

As described in the Introduction, the main interest in $NAD(P)^+$-dependent oxidoreductases is their applications in the synthesis of fine chemicals. These include the production of amino acids and hydroxy acids; the production of alcohols, aldehydes, and ketones; and the modification of steroids. Several comprehensive reviews on the subject have been presented elsewhere *(2,86,87)*. Pertinent examples of several of these processes, including a novel application of $NAD(P)^+$ oxidoreductases for energy applications, are discussed.

Amino acid synthesis is carried out by the reductive amination from an α-keto carboxylic acid and NH_4^+ and is catalyzed by an amino acid dehydrogenase. Most investigations have involved the synthesis of L-leucine, L-alanine or L-phenylalanine. The synthesis of L-leucine by reductive amination of α-ketoisocaproate was carried out in a membrane reactor that contained L-leucine dehydrogenase, PEG-NAD$^+$ (20,000), and a formate/FDH regeneration system. A rate of $1 kgL^{-1}$ day $^{-1}$ was achieved with a TTN of 80,000 after 90 days *(11)*. However, the periodic addition of enzyme and cosubstrate was necessary to maintain the rate during the experiment. A similar setup for the production of L-*tert*-leucine from trimethylpyruvate achieved a TTN of 125,000 in 60 days *(88)*.

The same reactor concept and regeneration system were used for the synthesis of L-phenylalanine from phenylpyruvate using L-phenylalanine dehydrogenase. A rate of production of phenylalanine of $456 gL^{-1}$ day^{-1} and a TTN of 600,000 in 350 h were reported *(89)*.

L-Alanine can be synthesized by the oxidation of either lactate or malate to pyruvate followed by reductive amination via alanine dehydrogenase. The oxidation of lactate provides the reducing equivalents for the regeneration of NADH. Because a separate cofactor regeneration system is not needed, this technique has been described as a coupled-substrate approach. L-Alanine was produced from lactate in a hollow-fiber reactor. The NAD$^+$ and substrates were fed into the reactor continuously *(90)*.

The rate of alanine production was 61.7 mmol L^{-1} h^{-1}, with a TTN of 4850 for the residence time of NAD^+ in the reactor. Malate/malate dehydrogenase can also be used to produce pyruvate. In a 100 mL batch reaction a conversion of 95% after 72 h and a TTN of 763 were obtained (91). L-Alanine can be directly synthesized from pyruvate; this system has been used to test dextran-bound NAD^+ derivatives (92).

Hydroxy acids can be produced from either the reduction of α-keto acids or the oxidation of alcohols. Hydroxy isocaproate dehydrogenase and LDH are used for the reduction of several keto acids. The continuous production of L-α-hydroxy isocaproate and L-trimethyl lactate was described in a membrane reactor where PEG-NAD^+ was regenerated by a formate/FDH system. Yields of 411 and 440 gL^{-1} day^{-1} and TTN values of 40,000 and 60,000, respectively, were obtained (88). The enzymatic conversion of pyruvate to D-lactate via LDH was used to test different cofactor regeneration schemes. The TTNs varied from 1500 in 15 days (93) using formate/FDH to 40,000 for GDH (*Bacillus cereus*) in 6 days (15).

GDH from *Gluconobacter scleriodes* was used for the regeneration of NADPH during the synthesis of L-leucovorin from dihydrofolate via dihydrofolate reductase in a stirred reactor (94).

The production of alcohols takes place by three main synthesis routes: the reduction of ketones, aldehydes, or sugars. The reduction of ketones via alcohol dehydrogenase is the most common method for preparing alcohols. Alcohol dehydrogenase from *Thermoanaerobium brockii* can be used to make aliphatic lactones (95). Isopropanol and *T. brockii* alcohol dehydrogenase can be used as an efficient NADPH regeneration system, and TTNs of 10,000 to 100,000 could be reached. Another system used the same enzyme to transform sulcatone to s-sulcatol in a continuous process (95,96). The cofactor, PEG-NAD^+, was retained by a charged ultrafiltration membrane. TTN values of 4500 were obtained. The use of a charged membrane gave better results than the use of an uncharged one.

L-menthol is a compound that is important in the synthesis of drugs and perfumes. Because it is poorly soluble, its enzymatic synthesis is carried out in a two-phase membrane bioreactor. The organic phase is separated from the aqueous phase by a microfiltration hydrophobic membrane. L-menthol was synthesized from L-menthone by menthone reductase (98). The cofactor was regenerated by the same enzyme using methyl isobutyl carbinol. For 270 h, production was 46.1 gL^{-1} day^{-1} and decreased by one-half after 607 h. The TTN reached 2500–3020 in 1 month. Another biphasic system, which was utilized by Grünwald et al. (99), also showed promise. Horse liver alcohol dehydrogenase and NAD^+ were fixed in the hydration layers at the surface of glass beads. The beads were suspended in organic solvent with the substrate and ethanol for regeneration. The reduction of 2-methyl valeraldehyde was achieved in this manner and optically active alcohol was produced.

The two main types of steroid functionalization, hydroxylation and dehydrogenation, are carried out by hydroxy steroid dehydrogenases (HSDHs). The oxidation of cholic acid to 12-ketochenodeoxycholic acid is carried out in the presence of 12-α–HSDH and $NADP^+$. α-Ketoglutarate and glutamate dehydrogenase were used as a regeneration system. A yield of 12 gL^{-1} in 50 h with a TTN of $NADP^+$

of 800 was obtained *(100)*. Immobilization of the enzymes on Sepharose CL-4B made it possible to increase the yield to 40 gL^{-1} in 4 days *(101)*.

A recent application for the utilization of NADP$^+$ in a biochemical pathway with cofactor recycling is the enzymatic conversion of sugars to molecular hydrogen *(102)*. This system is centered on two NADP$^+$-dependent oxidoreductases: GDH from *Thermoplasma acidophilum* and hydrogenase from *Pyrococcus furiosus*. A variety of monosaccharides and polysaccharides can serve as substrates in the reaction, which is depicted in Fig. 7. Glucose is the primary substrate for GDH. However, if other carbohydrate-degrading enzymes are included, additional raw materials such as cellulose, starch, lactose, and sucrose can also be used for the production of H$_2$ *(103,104)*. A 64% yield of H$_2$ from glucose has been demonstrated. This represents a TTN of 64 for recycling of the cofactor. The reactions are carried out at 55°C and pH 7.0. It has been shown that the NADP$^+$ is largely unstable under these reaction conditions. Furthermore, the rate of H$_2$ production can be restored by the addition of more NADP$^+$ Fig. 8).

Summary

The main prerequisite for the development of a biotechnological process is that it can be carried out efficiently and economically. In processes that require NAD(P)$^+$ dependent enzymes, the major costs are the enzyme and the NAD(P)$^+$. Advances in recombinant DNA technology and the development of high-level expression systems have the potential to cut the costs of enzyme production. In the case of the cofactor, efficient methods of regeneration and recovery are required.

High yields of products can be obtained in continuous-reaction systems that use macromolecular cofactor derivatives or in batch reactors that use the native cofactor. Recently, it has been demonstrated that the native coenzyme can be retained in a conventional ultrafiltration bioreactor by the addition of charged polymers to the reaction mixture. This technology may prove to be cost-effective for some applications in comparison with the expensive laborious synthesis of macromolecular derivatives.

Enzymatic methods continue to be the most effective techniques for the regeneration of cofactors. However, the fully integrated NAD$^+$ dehydrogenase electrode described by Bardea et al. *(81)* and the semisynthetic oxidases described by Yoma et al. *(64,65)* will certainly have applications for the regeneration of oxidized cofactors.

Acknowledgement

The authors are grateful to the Defense Advance Research Projects Agency for their support of this work, B. R. Evans and S. M. Davern for constructive criticisms, Marsha Savage for editorial services, and Laura Wagner for secretarial services.

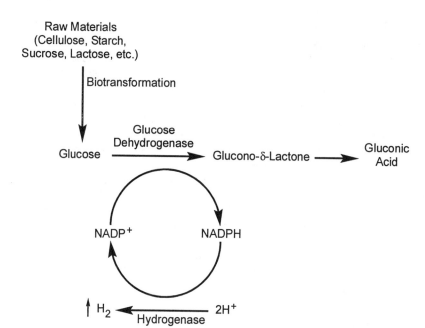

Figure 7. Pathway for the production of hydrogen from biomass.

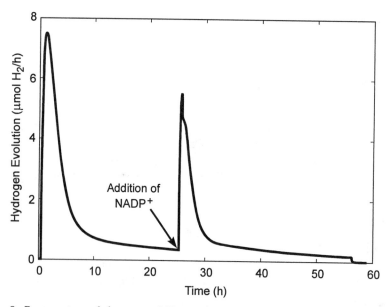

Figure 8. Restoration of the rate of H_2 production by the addition of NADP$^+$. The starting reaction mixture contained 50 mM glucose, 0.5 mM NADP$^+$, 5 U glucose dehydrogenase, and 32 U hydrogenase in 50 mM HEPES buffer, pH 7.0. Additional NADP$^+$, to a final concentration of 0.5 mM, was added after approximately 25 h.

References

1. Chenault, H. K.; Whitesides, G. M. *Appl. Biochem. Biotechnol.* **1987**, *14*, 147.
2. Bückmann, A. F.; Carrea, G. *Adv. Biochem. Eng.* **1989**, *39*, 97.
3. Nidetzky, B.; Haltrich, D.; Kuble, K. D. *Chemtechnol.* **1996**, *26*, 31.
4. Lowry, O. H.; Passonneau, J. V.; Rock, M. K. *J. Biol. Chem.* **1961**, *236*, 2756.
5. Young, S. R.; Bloom, A. D. *Biochem. Genet.* **1972**, *7*, 243.
6. Wu, J. T.; Wu, L. H.; Knight, J. A. *Clin. Chem.* **1986**, *32*, 314.
7. Rover, L., Jr.; Fernandes, J. C. B.; de Oliveria Neto, G.; Kubota, L. T.; Katekawa, E.; Serrano, S. H. P. *Anal. Biochem.* **1998**, *260*, 50.
8. Fawcett, C. P.; Ciotti, M. M.; Kaplan, N. O. *Biochim. Biophys. Acta* **1961**, *54*, 210.
9. Wenz, I.; Loesche, W.; Till, U.; Petermann, H.; Horn, A. *J. Chromatogr.* **1976**, *120*, 187.
10. Lee, L. G.; Whitesides, G. M. *J. Am. Chem. Soc.* **1985**, *107*, 6999.
11. Wichmann, R.; Wandery, C.; Bückmann, A. F.; Kula, M.-R. *Biotech. Bioeng.* **1981**, *23*, 2789.
12. Kroner, K. H.; Schütte, H.; Stach, W.; Kula, M. -R. *J. Chem. Technol. Biotechnol.* **1982**, *32*, 130.
13. Wong, C.-H.; Whitesides, G. M. *J. Am. Chem. Soc.* **1981**, *103*, 4890.
14. Hirschbein, B. L.; Whitesides, G. M. *J. Am. Chem. Soc.* **1982**, *104*, 4458.
15. Wong, C. 'H.; Dreuckhammer, D. G.; Sweers, H. M. *J. Am. Chem. Soc.* **1985**, *107*, 4028.
16. Pollak, M.; Blumenfeld, H.; Wax, M.; Baughn, R. L.; Whitesides, G. M. *J. Am. Chem. Soc.* **1980**, *102*, 6324.
17. Kulbe, K. D.; Schwab, U.; Howaldt, M. *Ann. N.Y. Acad. Sci.* **1987**, *501*, 216.
18. Everse, J.; Zoll, E. L.; Kahan, L.; Kaplan, N. O. *Bioorg. Chem.* **1971**, *1*, 207.
19. Anderson, B. M. In *Pyridine Nucleotide Coenzymes*; Everse, J.; Anderson, B.; You, K., Eds.; Academic Press: New York, 1982; pp. 91.
20. Adams, M. J.; Buehner, M.; Chandrasekhar, K.; Ford, G. C.; Hackert, M. L.; Kaplan, N. O.; Tayler, S. S. *Proc. Nat. Acad. Sci. USA* **1973**, *70*, 1968.
21. Månsson, M. O; Mosbach, K. In *Coenzymes and Cofactors*, Dolphin, D.; Paulson, R.; Avramovic, O., Eds.; Wiley: New York, 1987; *2B*, 217.
22. Bückmann, A. F.; Wray, V. *Biotechnol. Appl. Biochem.* **1992**, *15*, 303.
23. Lowe, C. R. In *Topics in Enzyme and Fermentation Biotechnology*; Wiseman, A., Ed.; Ellis Horwood Ltd: Chichester, UK, 1978; pp. 13.
24. Lowe, C.R.; Mosbach, K. *Eur. J. Biochem.* **1974**, *49*, 511.
25. Murumatsu, M.; Urabe, I.; Yamada, Y.; Okada, H. *Eur. J. Biochem.* **1997**, *80*, 111.
26. Sakamoto, H.; Nakamura, A.; Urabe, Y.; Okada, H. *J. Ferment. Technol.* **1986**, *6*, 511.
27. Kishimoto, T.; Itami, M.; Yomo, T.; Urabe, I.; Yamada, Y.; Okada, H. *J. Ferment. Bioeng.* **1991**, *71*, 447.
28. Zappelli, P.; Rossodivita, A.; Re, L. *Eur. J. Biochem.*, **1975**, *54*, 475.
29. Zappelli, P. Pappa, R.; Rossodivita, A.; Re, L. *Eur. J. Biochem.* **1977**, *72*, 309.
30. Bückmann A. F. *Heterocycles* **1988**, *27*, 1623.

31. Weibel, M. K.; Fuller, C. W.; Stadel, J. M.; Bückmann, A. F.; Doyle, T.; Bright, H. J. In *Enzyme Engineering*; Pye, E. K.; Wingard, L. B., Jr., Eds.; Plenum Press: New York, **1974**; *Vol. 2*, pp. 203.
32. Schmidt, F.-L.; Grenner, G. *Eur. J. Biochem.* **1976**, *67*, 295.
33. Grenner, G.; Schmidt, H.-L.; Volkl, W. *Hoppes-Seyler's Z. Physiol. Chem.* **1976**, *357*, 887.
34. Lindberg, M.; Larsson, P.-O.; Mosbach, K. *Eur. J. Biochem.* **1973**, *40*, 187.
35. Lee, C.-Y; Lappi, D. A.; Wermuth, B.; Everse, J.; Kaplan, N. O. *Arch. Biochem. Biophys.* **1974**, *163*, 561.
36. Lee, C.-Y.; Kaplan, N. O. *Arch. Biochem. Biophys.* **1975**, *168*, 665.
37. Zappelli, P.; Rossodivita, A.; Prosperi, G.; Pappa, R.; Re, L. *Eur. J. Biochem.* **1976**, *62*, 211.
38. Zappelli, P.; Pappa, R.; Rossodivita, A.; Re, L. *Eur. J. Biochem.* **1978**, *89*, 491.
39. Yamazaki, Y.; Maeda, H.; Suzuki, H. *Biotechnol. Bioeng.* **1976**, *18*, 1761.
40. Sakaguchi, Y.; Murachi, T. *J. Appl. Biochem.* **1980**, *2*, 117.
41. Malinauskas, A. A.; Kulys, J. J. *Biotechnol. Bioeng.* **1978**, *20*, 769.
42. Bückmann, A. F.; Kula, M. R.; Wichmann, R.; Wandrey, C. *J. Appl. Biochem.* **1981**, *3*, 301.
43. Okuda, K.; Urabe, I.; Okada, H. *Eur. J. Biochem.* **1985**, *151*, 33.
44. Bückmann, A. F. *Biocatalysis* **1987**, *1*, 173.
45. Axen, R.; Porath, J.; Ernback, S. *Nature* **1967**, *214*, 1302.
46. Yoshikawa, M.; Goto, M.; Ikura, K.; Sasaki, R.; Chiba, H. *Agric. Biol. Chem.* **1983**, *46*, 207.
47. Sogin, D. C. *J. Neurochem.* **1976**, *27*, 1333.
48. Bückmann, A. F.; Morr, M.; Johansson, G. *Macromol. Chem.* **1981**, *182*, 1379.
49. Fuller, C. W.; Rubin, J. R.; Bright, H. J. *Eur. J. Biochem.* **1980**, *103*, 421.
50. Nakamura Y.; Suye, S.; Kira, J.; Tera, H.; Tabata, I.; Senda, M. *Biochim. Biophys. Acta* **1996**, *1289*, 221.
51. Aizawa, M.; Coughlin, R. W.; Charles, M. *Biochim. Biophys. Acta* **1975**, *385*, 362.
52. Persson, M.; Månsson, M.-O.; Bülow, L.; Mosbach, K. *Bio/technology* **1991**, *9*, 280.
53. Holbrook, S. R.; Muskal, M.; Kim, S.-H. *Protein Eng.* **1990**, *3*, 659.
54. Månsson, M. O.; Larsson, P. O.; Mosbach, K. *Eur. J. Biochem.* **1987**, *86*, 455.
55. Månsson, M. O.; Larsson, P. O.; Mosbach, K. *FEBS Lett.* **1979**, *98*, 309.
56. Gacesa, P.; Venn, R. F. *Biochem. J.* **1979**, *177*, 369.
57. Kovar, J.; Simek, K.; Kucera, I.; Matyska, L. *Eur. J. Biochem.* **1984**, *139*, 585.
58. Schäfer, H. G.; Jacobi, T.; Eichorn, H.; Woenckhaus, C. *Biol. Chem. Hoppe-Seyler* **1986**, *367*, 969.
59. Eguchi, T.; Iizuka, T.; Kagotami, T.; Lee, J. H.; Urabe, I.; Okada, H. *Eur. J. Biochem.* **1986**, *155*, 415.
60. Nakamura, A.; Urabe, I.; Okada, H. *J. Biol. Chem.* **1986**, *261*, 16792.
61. Kato, N.; Yamagami, T.; Shimao, M.; Sakazawa, C. *Appl. Microb. Biotechnol.* **1987**, *25*, 415.
62. Goulas, P. *Eur. J. Biochem.* **1987**, *168*, 469.

63. Woenckhaus, C.; Koob, R.; Burkhard, A.; Schäfer, H. G. *Bioorg. Chem.* **1983**, *12*, 45.
64. Yomo, T.; Urabe, I.; Okada, H. *Eur. J. Biochem.* **1991**, *200*, 759.
65. Yomo, T.; Urabe, I.; Okada, H. *Eur. J. Biochem.* **1992**, *203*, 533.
66. Blaedal, W. J.; Jenkins, R. A. *Anal. Chem.* **1975**, *47*, 1337.
67. Kelly, R. M.; Kirwain, D. *J. Biotechnol. Bioeng.* **1977**, *19*, 1215.
68. Jaegfeldt, H.; Torstensson, A. B. C.; Johansson, G. *Anal. Chim. Acta* **1981**, *97*, 221.
69. Jenson, M. A.; Elving, P. J. *Biochim. Biophys. Acta* **1984**, *764*, 310.
70. Elving, P. J.; Schmakel, C. O.; Santhanam, K. S. *Crit. Rev. Anal. Chem.* **1976**, *6*, 1.
71. Huck, H.; Schmidt, H.-L. *Angew. Chem., Int. Ed. Engl.* **1981**, *20*, 402.
72. Tse, D.; Kuwana, T. *Anal. Chem.* **1978**, *50*, 1315.
73. Gorton, L ; Tortensson, A.; Jaegfeld, H.; Johansson, G. *J. Electroanal. Chem.* **1984**, *161* 103.
74. Gorton, L.; Johansson, G.; Tortensson, A. *J. Electroanal. Chem.* **1985**, *196*, 81.
75. Gorton, L. *J. Chem. Soc., Faraday Trans. I* **1986**, *82*, 1245.
76. Malinauskas, A.; Kulys, J. *J. Anal. Chim. Acta* **1978**, *98*, 31.
77. Torstensson, A.; Gorton, L. *J. Electroanal. Chem.* **1981**, *130*, 199.
78. Karyakin, A. A.; Bobrova, O. K.; Karyakina, E. E. *J. Electroanal. Chem.* **1995**, *399*, 179.
79. Weinkamp, R.; Steckhan, E. *Angew. Chem., Int. Ed. Engl.* **1982**, *21*, 782.
80. Lo, H. C.; Buriez, O.; Kerr, J. B.; Fish, R. H. *Angew. Chem., Int. Ed.* **1999**, *38*, 1429.
81. Bardea, A.; Katz, E.; Bückmann, A. F.; Willner, I. *J. Am. Chem. Soc.* **1997**, *119*, 9114.
82. Wandrey, C.; Wichmann, R. In *Enzymes and Immobilized Cells in Biotechnology;*. Laskin, A., Ed.; Addison Wesley, Reading, MA, 1985; pp. 177.
83. Ikemi, M.; Ishimatsu, Y.; Kise, S. *Biotechnol. Bioeng.* **1990**, *36*, 155.
84. Obón, J. M.; Manjón, A.; Iborra, J. L. *Biotechnol. Bioeng.* **1997**, *57*, 510.
85. Obón, J. M.; Almagro, M. J.; Manjón, A.; Iborra, J. L. *J. Biotechnol.* **1996**, *50*, 27.
86. Devaux-Basseguy, R.; Bergel, A.; Comtat, M. *Enzyme Microb. Technol.* **1997**, *20*, 248.
87. Aldercreutz, P. *Biocatal. Biotrans.* **1996**, *14*, 1.
88. Wandrey, C.; Bossow, B. *Int. Conf. Chem. Biotechnol. Biol. Act. Proc. 3rd*, **1987**, *1*, 195.
89. Hummel, W.; Schütte, H.; Schmidt, E.; Wandrey, C.; Kula, M.-R. *Appl. Microbiol. Biotechnol.* **1987**, *26*, 409.
90. Fujii, T.; Miyawaki, O.; Yano, T. *Biotechnol. Bioeng.* **1991**, *38*, 1166.
91. Suye, S-I.; Kaweagoe, M.; Inuta, S. *Can. J. Chem. Eng.* **1992**, *70*, 306.
92. Davies, P.; Mosbach, K., *Biochim et Biophys. Acta*, **1974**, *3670*, 329.
93. Shaked, Z.; Whitesides, G. M. *J. Am. Chem. Soc.* **1980**, *102*, 7104.
94. Eguchi, T.; Kuge, Y.; Inoue, K.; Yoshikawa, N.; Mochida, K.; Uwajima, T. *Biosci. Biotechnol. Biochem.* **1992**, *56*, 701.
95. Keinan, E.; Seth, K. K.; Lamed, R. *Ann. N.Y. Acad. Sci.* **1987**, *501*, 130.

96. Röthig, T. R.; Kuble, K. D.; Bückmann, A. F.; Carrea, G. *Biotechnol. Lett.* **1990**, *12*, 353.

97. Kulbe, K. D.; Howaldt, M. W.; Schmidt, K.; Röthig, T. R.; Chmiel, H. *Ann. N.Y. Acad. Sci.* **1990**, *623*, 820.

98. Kise, S.; Hayashida, M. *J. Biotechnol.* **1990**, *14*, 221.

99. Grünwald, J.; Wirz, B.; Scollar, M.P.; Klibanov, A. M. *J. Am. Chem. Soc.* **1986**, *108*, 6732.

100. Carrea, G.; Bovara, R.; Cremonesi, O.; Lordi, R. *Biotechnol. Bioeng.* **1984**, *26*, 560.

101. Carrea, G.; Bovara, R.; Longhi, R.; Riva, S. *Enzyme Microb. Technol.* **1985**, *7*, 597.

102. Woodward, J.; Mattingly, S. M.; Danson, M.; Hough, D.; Ward, N.; Adams, M. *Nature Biotechnol.* **1996**, *14*, 872.

103. Woodward, J.; Orr, M. *Biotechnol. Prog.* **1998**, *14*, 897

104. Woodward, J.; Cordray, K. A.; Edmonston, R. J.; Blanco-Rivera, M.; Mattingly, S. M.; Evans, B. R. *Energy Fuels* **2000**, in press.

Chapter 9

Production of Galacto-Oligosaccharides from Lactose by Immobilized β-Galactosidase

Shang-Tian Yang and Julia A. Bednarcik

Department of Chemical Engineering, The Ohio State University, Columbus, OH 43210

Galacto-oligosaccharides (GOS) and oligosaccharides in general have received a lot of attention recently, mainly due to their many beneficial health effects and wide applications as prebiotic food. Production of GOS from lactose by enzyme reaction is reviewed in this paper. The enzyme β-galactosidase can be used to produce GOS containing 2 to 5 galactose units and one glucose unit from lactose. Depending on the enzyme source and reaction conditions, the GOS yield varied from below 20% to as high as 67% (w/w). In general, a higher initial lactose concentration gave a wider range of GOS types produced and increased GOS yield. Reactions in an organic solvent did not increase GOS production, but rapidly inactivated the enzyme. Using a commercial enzyme from *Aspergillus oryzae*, a maximum GOS yield of ~71%, based on lactose reacted, was obtained at low lactose conversions (~10%), but the yield decreased with increasing lactose conversion. An integrated immobilized enzyme reactor-separator process, which continuously removes GOS from the reaction media, would give the highest possible GOS yield (>65%) from lactose. Effects of enzyme immobilization and methods to separate GOS, such as nanofiltration, are also discussed in this article.

Introduction

In this paper, a brief review on what GOS are, their health benefits, current and future market situations, and current production technologies using microbial β-

galactosidase is provided. The enzyme reaction kinetics and factors affecting GOS production from lactose are examined in details. A novel process for optimizing GOS production from lactose is also discussed in this article.

Oligosaccharides

Oligosaccharides (OS) are large sugar molecules containing three to twenty monosaccharide units, joined together by glycosidic bonds (1,2). These short chains are usually linear, but also come in branched forms. There are approximately eighty types of glycosidic bonds naturally found, with the monosaccharides mannose, N-acetylglucosamine, N-acetylmuramic acid, glucose, 6-deoxygalactose, galactose, N-acetylneuraminic acid and N-acetylgalactosamine being the most often used in nature. Some oligosaccharides naturally occur in the human body, and they mediate a variety of intercellular interactions and affect protein and lipid properties. Table I lists major food-grade oligosaccharides along with their chemical sources and market sizes (2). These oligosaccharides are not naturally produced in the human body, but were found to have numerous physiological benefits for mammals (3-5). Thus, these oligosaccharides are of commercial interest and their production and applications (such as in prebiotic foods for specified health use) are therefore increasing rapidly.

Table I. Various Types of Food-Grade Oligosaccharides

Class of Oligosaccharide	Chemical Sources	Production in 1995 (t)
Galacto-oligosaccharides (GOS)	Lactose	15000
Lactulose	Lactose	20000
Lactosucrose	Lactose + Sucrose	1600
Fructo-oligosaccharides (FOS)	Inulin, Sucrose	12000
Palatinose (isomaltulose) OS	Sucrose	5000
Glucosyl sucrose	Sucrose + Maltose	4000
Malto-oligosaccharides (MOS)	Starch	10000
Isomalto-oligosaccharides	Starch	11000
Gentio-oligosaccharides	Starch	400
Cyclodextrins	Starch	4000
Soybean oligosaccharides	Raffinose, Stachyose	2000
Xylo-oligosaccharides	Xylan	300

SOURCE: adapted from Reference 2.

Food-grade OS can be broadly classified by their source, such as OS derived from lactose, soybean, starch, or sucrose. Oligosaccharides are also classified according to their main monosaccharide component: FOS composed of fructose and glucose, or MOS composed of glucose. Oligosaccharides in a specific group have a few variations in their chain length (trisaccharides, tetrasaccharides, etc.) and linkage type (e.g., β–1-4, α–1-6, and β–1-6 linkages), depending on the sources and production methods used. It should be noted that lactulose is a disaccharide and many food-grade oligosaccharides also contain disaccharides. Both OS and these disaccharides have similar functions and health benefits shown in Table II.

Table II. Physiological Functions and Health Benefits of Oligosaccharides

Physiological Functions and Health Benefits	
Non-digestibility	• Not digested by stomach enzymes, but pass to the intestines. Do not enter the bloodstream, do not enter the pancreas or cause the secretion of insulin, thus can be used as a "natural" sweetener for diabetes patients. • Can be used as low-calorie sugars, with a sweetness of 20% to 50% of sucrose. • Similar to dietary fibers; can improve constipation
Low cariogenicity	• Not utilized by oral streptococcus (*S. mutans*) which produces insoluble glucans, promoting dental caries.
Growth promotion of bifidobacteria	• Reduction of detrimental bacteria in intestinal microflora • Production of nutrients (B vitamins); increasing calcium resorption • Decrease of toxic products and resulting effects • Regulation of Bowels • Reduction of cholesterol and blood pressure • Reduction of age-related ailments

Health Benefits

The various physiological benefits that mammals obtain from oligosaccharides, including GOS, are received via two different routes: direct (through oligosaccharide ingestion) and indirect (human ingestion then digestion by beneficial bifidobacteria that reside in the colon). Table II shows the main physiological functions and health benefits of oligosaccharides. Oligosaccharides selectively increase desirable intestinal bacteria, such as *Bifidobacteria* and *Lactobacilli*, and decrease undesirable bacteria, such as *Bacteroideceae*. Thus, they help maintain healthy intestinal bacterial flora and provide the health benefits listed in Table II due to a high *Bifidobacteria* flora (*6-13*). In addition, ingestion of these oligosaccharides also may improve constipation (a contributor to cancer development in the large intestine), hyperammonemia, endotoxemia and symptoms of non-compensatory liver cirrhosis.

Galacto-oligosaccharides (GOS)

In this paper, we focus on GOS since it has large commercial interest and is currently produced from lactose in an enzymatic reaction catalyzed by microbial β–galactosidase. GOS are natural constituents in human milk (*14*) as well as in foods such as garlic and onions. Food grade GOS are commercially produced from lactose and contain various oligosaccharides. Their chemical formula is $(Galactose)_n$-Glucose, with n ranging from 2 to 4. The galactose-galactose linkage is a β–(1→3), β–(1→4), or β–(1→6) linkage, with the β–(1→4) linkage being predominant; the galactose-glucose linkage is mainly β–(1→4). Some disaccharides (e.g., allolactose and galactobiose) are also present in GOS. GOS are specifically known as "Bifidus growth factor" since they stimulate the growth of *Bifidobacteria* in the human body (*15-18*).

Market and Applications

Four companies, three located in Japan and one in the Netherlands, currently produce GOS. In 1995, approximately 15,000 tons of GOS were produced (*2*). In Japan, OS are already available through addition to over 450 foods. In the United States and North America, oligosaccharides also should have many marketable areas since Americans are demanding "natural" foods with beneficial health effects. Some such areas include:

- Addition to processed foods, including beverages (milk, soft drinks), frozen foods, and snacks such as cookies and chips.
- Use as a dietary supplement. A limited number of fructo-oligosaccharides are already available in the U.S. as a herbal dietary supplement, but their use is unregulated and their health effects unapproved by the FDA.
- Inclusion in infant formulas (specifically GOS) to make them more closely resemble mother's breast milk.
- Addition to animal feeds. This will help reduce disease and keep the animals healthy, and it is safer than the current methods of adding antibiotics to feeds (since it does not promote antibiotic-resistant microorganisms); oligosaccharides are currently being added to feeds in Japan.
- Low-calorie sweetener. This is especially useful for diabetics and obese patients, since the sugars are not digested but taste and sweetness are not sacrificed.
- New drug development, including cell adhesion or transplantation based drugs.
- Other non-food applications including cosmetic use and for mouthwashes.

Production Methods

There are four main oligosaccharide production methods currently in use. Oligosaccharides "produced" in North America and Europe are usually by extraction and purification from natural sources. One example is Raftilose, a mixture containing up to 95% fructo-oligosaccharides and monosaccharides, produced from chicory root in Belgium. The main difficulty with this production method is in regulating and controlling the oligosaccharide content in the vegetable feed. Raffinose and stachyose are two other oligosaccharides extracted from plants, including soybeans and beets. A process such as this one would be unable to gain FDA approval for marketing in the U.S. due to variations in the oligosaccharide source and the resulting variations in the oligosaccharide product. Another production method uses cell cultures to create oligosaccharide chains. Cheil Foods & Chemicals (Seoul, Korea) uses immobilized *Aureobasidium pullulans* cells to produce fructo-oligosaccharide syrups from sucrose (*19*). However, this method is not widely applied, since production costs are high due to a low oligosaccharide yield and expensive product recovery.

Another popular method to create oligosaccharides is by breaking long polysaccharide chains through the use of enzymes or acid addition. This method is used to produce xylo-oligosaccharides and agar-oligosaccharides through hydrolysis. The downside of this technique is that it produces oligosaccharides with a variety of chain lengths since hydrolysis may be difficult to control, giving an inconsistent product. The fourth method to produce oligosaccharides is via an enzymatic reaction from a low-cost substrate. Fructo-oligosaccharides, isomalturose, cyclodextrins, glucosyl sucrose, and GOS can all be created in this manner (*20*).

Enzymatic Reaction for GOS Production

The reactions for GOS synthesis have been studied using enzymes derived from various sources, but the reaction kinetics have not been deeply investigated or optimized for use in a continuous production process. The current technical literature is reviewed here, especially relating to the reactions for GOS synthesis and enzyme immobilization. Ways that one could improve the reaction, particularly in GOS yield, such that it would be suitable for an economically-feasible continuous production process are also discussed.

Reactions of β–Galactosidase

β–Galactosidase or lactase (EC 3.2.1.23) is one of the first of the class of oligosaccharide-hydrolyzing enzymes isolated and purified from various organisms, including plants, fungi, yeasts, bacteria, and animal organs. The enzyme lactase has long been used to hydrolyze lactose in the manufacture of some dairy products (*21-23*). The normal function of the β–galactosidase enzyme is to hydrolyze lactose to glucose and galactose. However, under certain reaction conditions, the same enzyme also catalyzes a transgalactosylation reaction, which leads to the production of oligosaccharides (*24-27*).

Lactase has been studied extensively from the standpoint of its reaction mechanism, in both glycosidase and glycotransferase actions (*25*). The biological and natural substrate for β–galactosidase is lactose. Lactose, however, is not the only substrate, and not always the best. The enzyme catalyzes the hydrolysis of β–galactosidic linkages such as those found in lactose. The active site of lactase has been proposed to be an imidazole and a sulfhydryl group. The sulfhydryl group acts as a general acid to protonate the galactosidic oxygen atom and the imidazole group acts as a nucleophile that attacks the nucleophilic center at the first carbon of the galactose molecule. The glycosidases are also known to catalyze transfer reactions;

136

that is, the sugar residue forming the glycan part of the substrate molecule may be transferred to a sugar or to some hydroxyl acceptor to form an oligosaccharide. Up to twenty different GOS have been found, primarily trisaccharides and disaccharides, but tetrasaccharides, pentasaccharides and even larger ones have also been reported (*1,26*).

The reaction kinetics has been well characterized for lactase from various sources. The enzymatic hydrolysis of lactose follows the Michaelis-Menten kinetics with competitive inhibition by galactose (*28*). A reversible reaction kinetic model and a more complex model involving the mutarotation of galactose which results in a stronger competitive inhibitor, α-galactose, have also been proposed (*29*). Other models that also describe the transgalactosylation reactions also exist (*30*). An extended Michaelis-Menten model with trisaccharide intermediates has also been used to model the reactions (*25*). A general reaction scheme is shown in Table III.

Table III. Reactions for GOS Production from Lactose by β-galactosidase.

Reactions	Symbols
$L + E + H_2O \leftrightarrow (E\text{-}L) \rightarrow (E\text{-}Gal) + Glu$	L: lactose
$(E\text{-}Gal) \leftrightarrow E + Gal$	E: β-galactosidase
	Gal: Galactose
$(E\text{-}Gal) + L \leftrightarrow (E\text{-}O1) \leftrightarrow E + O1 + H_2O$	Glu: Glucose
$(E\text{-}Gal) + O1 \leftrightarrow (E\text{-}O2) \leftrightarrow E + O2 + H_2O$	O1: trisaccharide
	O2: tetrasaccharide
$(E\text{-}Gal) + O2 \leftrightarrow (E\text{-}O3) \leftrightarrow E + O3 + H_2O$	O3: pentasaccharide
$Gal + (E\text{-}Gal) \leftrightarrow (E\text{-}GB) \leftrightarrow E + GB + H_2O$	GB: galactobiose
	AL: allolactose
$Glu + (E\text{-}Gal) \leftrightarrow (E\text{-}AL) \leftrightarrow E + AL + H_2O$	(E-X): Enzyme-X complex

SOURCE: adapted from Reference 27.

Due to the dual catalytic activities of β-galactosidase, the enzyme is involved in competing reactions when incubated with lactose. In general, transgalactosylation dominates early on in the reaction, producing GOS with a high yield. As the lactose conversion increases, the enzyme's hydrolytic activity then takes over; the final products after 100% lactose conversion are glucose and galactose. Galactose is a competitive inhibitor, causing the enzyme's actions to greatly slow down at high lactose conversions, reducing the rate of the reaction. Figure 1 shows typical batch kinetics for lactose hydrolysis and oligosaccharide formation catalyzed by a commercial lactase enzyme derived from *Aspergillus oryzae*. In this figure, disaccharides such as allolactose and galactobiose are not separated from lactose and they are not considered as part of the GOS.

The catalytic activity of lactase is affected by temperature, ionic strength, pH, inhibitors, and substrate concentration. These effects and some other properties are quite different for lactase from different sources. Since the reaction rate is dependent on the enzyme activity, substrate concentration, and other conditions, it is difficult to compare the enzyme reaction kinetics using the time course data shown in Figure 1.

On the other hand, the GOS production kinetics is closely related to lactose conversion, as shown in Figure 2, which is usually only affected by the initial lactose concentration and enzyme sources. As can be seen in Figure 2, GOS production increases with increasing lactose conversion until a maximum is reached at ~50% conversion; the amount of GOS then decreases as the reaction continues, and they eventually disappear completely at 100% conversion.

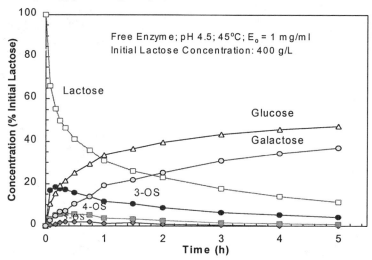

Figure 1. *Batch reaction kinetics of lactose hydrolysis and oligosaccharides formation catalyzed by β-galactosidase from Aspergillus oryzae. (All disaccharides are reported as lactose here since they were not separated in the experiment).*

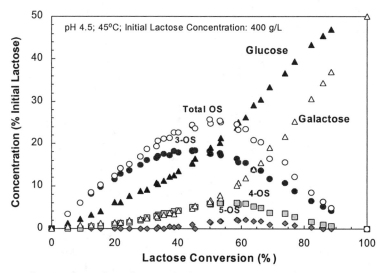

Figure 2. *Products from lactose hydrolysis and transgalactosylation catalyzed by Aspergillus oryzae β-galactosidase as affected by lactose conversion.*

Table IV. GOS Production from Lactose with Microbial β-galactosidase.

Enzyme Sources	Reaction Conditions			Max. GOS (w %)	Reference (31-51)
	Conc. (g/L)	T (°C)	pH		
Aspergillus niger	200	45	4.5	18.9	Kim et al., 1990
Aspergillus oryzae	400	40	4.5	32	Iwasaki et al., 1996
Bacillus circulans	456	40	6.0	40	Mozaffar et al., 1986
Bacillus subtilis	200	10	7.0	18	Rahim & Lee, 1991
Bullera singularis	100	45	4.8	55	Shin et al., 1998
Caldocellum saccharolyticum	700	80	6.3	42	Stevenson et al., 1996
Cryptococcus laurentii	100			47	Ohtsuka et al., 1988
Kluyveromyces fragilis	350	35	6.2	45	Shukla, 1975
Kluyveromyces lactis	200	45	7.0	31	Foda & Lopez-Leiva, 1999
Penicillium simplicissimum	600	50	6.5	30.5	Cruz et al., 1999
Pyrococcus furiosus[1]	450	75	5.0	29	Boon et al., 1999
Rhodotorula minuta	360	60	6.0	64	Onishi & Yokozeki, 1996
Saccharomyces fragilis	350	35	6.2	45	Roberts & Pettinati, 1957
Saccharomyces lactis	200	35	6.2	12.5	Burvall et al., 1979
Saccharopolyspora rectivirgula	600	70	7.0	41	Nakao et al., 1994
Sirobasidium magnum	360	60	6.0	67	Onishi et al., 1996
Sporobolomyces singularis	100			50	Gorin et al., 1964
Streptococcus thermophilus	100	37	7	25	Greenberg & Mahoney, 1983
Sterigmatomyces elviae	360	60	4.5	60	Onishi & Tanaka, 1998
Thermus aquaticus	160	70	4.6	35	Berger et al., 1995
Trichoderma harzianum	150	30	7	32	Prakash et al., 1987

1. The enzyme is β-glucosidase

GOS Yield

One major consideration in developing a GOS production process is its yield. There are two ways for comparing the extent of transgalactosylation for GOS synthesis or GOS yield: the first is the process yield based on the GOS content in the final reaction mixture, and the second is the kinetic yield based on GOS produced per lactose reacted during the reaction. GOS content in the product mixture (as percent of total saccharides or initial lactose concentration) is usually reported in the literature and used to evaluate the transgalactosylation ability of an enzyme. Table IV lists some microbial β-galactosidases and the reaction conditions used for GOS production. Regardless of enzyme source, the maximum GOS content (or process yield) is usually obtained at a lactose conversion higher than 50% and is typically 20%-40%. The higher GOS yields (>50%) are usually those also include the disaccharides. Some OS, especially disaccharides, are formed from intramolecular transfer reactions. In this case, the OS are not degraded by the enzymes, such as β-galactosidase from *S. elviae*, very much as the reaction proceeds, resulting in a high OS yield with a high fraction of disaccharides.

A second, and perhaps more accurate, method of measuring an enzyme's transgalactosylation potential is by comparing the actual GOS yield from lactose reacted. Figure 3 shows the product yields in the enzyme reaction as a function of lactose conversion. More trisaccharides are formed at a lower conversion, and as reaction continues, less trisaccharides but more monosaccharides (glucose and galactose) are formed from the lactose reacted. Tetrasaccharide yield increases with increasing conversion and peaks at ~50% conversion, but then decreases as the reaction continues. Pentasaccharides follow a similar trend but peak at a higher conversion (~60%). Since trisaccharides constitute the majority of GOS produced in the reaction, total GOS yield follows a trend similar to that for trisaccharides. The theoretical maximum GOS yield is 75% based on trisaccharide formation. For longer GOS, the yield decreases and asymptotically approaches 50%. It should be mentioned that disaccharides (e.g., allolactose and galactobiose) are not considered in the total GOS yield calculation shown in Figure 3. If the disaccharides are also considered, the total GOS yield would be much higher since significant amounts of disaccharides can be formed from partial hydrolysis of trisaccharides and intramolecular bond transfer within lactose (*35*).

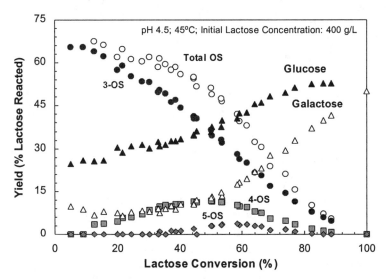

Figure 3. *Product yields from lactose reacted as affected by lactose conversion.*

Factors Affecting GOS Production

Enzyme Source

Enzymes from different sources have different characteristics: pH optima, temperature optima, and type and distribution of GOS. The results of GOS production from lactose using different sources of β-galactosidase are also quite different (see Table IV). Due to the different processing conditions, this wide variety of enzyme characteristics can be advantageous. For example, β-galactosidases from *Thermus Aquaticus* YT-1 and *Sterigmatomyces elviae* CBS8119 are highly thermostable, with temperature optima of 70°C (*52*) and 85°C (*53,54*), respectively. One consideration concerning the potential for industrial use is that β-galactosidase from most of the sources are not commercially available, not available in large quantities, or that which is available is not approved for food use.

The enzyme used in our work is the β-galactosidase derived from the mold *Aspergillus oryzae*. This enzyme has a pH optima of 4.5, and a temperature optima of 45°C (*55*). The lactase used in industry are usually from *Aspergillus niger* or *A. oryzae* because of their acidic pH optima (pH 3.5-4.5) and relatively high temperature stability. Since a typical substrate for an industrial production process would be acid whey (containing lactose), the low pH used by *A. oryzae* is desirable. Also, this enzyme is already used in food products, is commercially available, and relatively inexpensive compared to β-galactosidase from alternative sources. The only drawback of this enzyme is the maximum GOS content obtained is usually lower than 30%, which is far lower than the values reported for many bacterial enzymes.

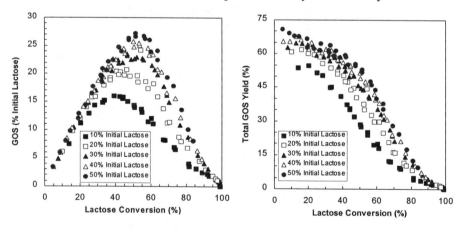

Figure 4. Effects of lactose concentration and conversion on GOS production. The experimental conditions were the same as those shown in Figure 1, except that the initial lactose concentration varied from 10% (100 g/L) to 50% (500 g/L).

Lactose Concentration

The production of oligosaccharides increases with the initial lactose concentration. As shown in Figure 4, GOS yields are higher with a higher initial lactose concentration at all lactose conversion levels. As the initial lactose concentration increases from 100 g/L to 500 g/L, the maximum GOS content in the product increases from 17% (at 40% conversion) to 28% (at 52% conversion). The maximum GOS yield from lactose reacted also increases with the initial lactose concentration, from 55% at 20% conversion to 71% at 5% conversion. It is noted that the effects are more prominent on larger oligosaccharides. As shown in Figure 5, as the initial lactose concentration increases from 100 g/L to 500 g/L, the trisaccharides content increases 38% (from 13% to 18%), while tetrasaccharides and pentasaccharides increase by 2.5-fold (from 2.6% to 6.5% and 1 % to 2.5%, respectively).

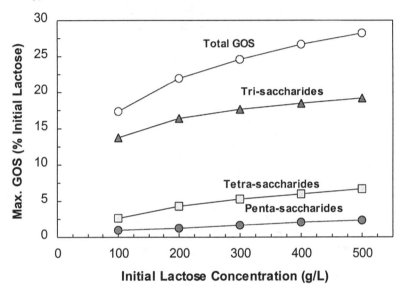

Figure 5. *Effects of initial lactose concentration on GOS production. The experimental conditions were the same as those in Figure 4.*

Compared to lactose concentration and lactose conversion, other process parameters such as enzyme concentration, pH and temperature have minimal effects on GOS production (*56*), although they all have profound effects on the reaction rates. The effects of enzyme immobilization are discussed in a separate section later in this article.

Methods to Increase GOS Synthesis

The oligosaccharide yield can be increased by shifting the equilibrium between the two competing reactions. Several methods have been proposed to shift the actions of β-galactosidase to favor transgalactosylation over lactose hydrolysis (*57*), including 1) reaction in non-aqueous media, 2) using high lactose (substrate) concentration, 3) adding high concentration of galactose, and 4) removing oligosaccharides from the medium. The effects of lactose concentration on GOS formation have been discussed in the previous section. Here we will discuss the other three methods.

Reactions in Organic Media

One method employed to increase GOS yields is the use of organic solvents. In general, water is essential to enzyme catalysis, allowing the enzyme to retain its native configuration and activity. However, it has been shown that only a small amount of water is necessary - as little as one monolayer of water surrounding the enzyme is enough to retain activity (*59,60*). Many enzymatic reactions are favorably altered by conducting them in an organic solvent – aqueous mixture (*61-63*), creating a micro-aqueous environment for the enzyme. For the hydrolysis of lactose, β-galactosidase adds a molecule of water to split the glycosidic bond between glucose and galactose. By reducing the amount of water available for hydrolysis, theoretically the hydrolytic reaction should be partially inhibited and the transgalactosylation reaction should then be favored (since water is released, not consumed, during transgalactosylation). In organic solvents containing a minimum amount of water the hydrolysis function is almost totally inhibited, and the catalysis of glycosidation reactions in organic solvent showed enhanced oligosaccharide synthesis (*64-66*). Shin and Yang (*67*) tested the production of GOS with β-galactosidase *of B. circulans* in six common organic solvents. They reported that the oligosaccharides yield increased from 38% (w/w) in aqueous media to 45% (w/w) in a 95% cyclohexane / 5% water mixture. It has also been observed that the presence of organic solvents could affect the type of oligosaccharides produced (*68*).

However, our recent study (*69*) with *A. oryzae* β-galactosidase showed that at the same lactose conversion there was no increase in GOS production for reactions in organic media (95% solvent, 5% water). Stevenson et al. (*70*) also reported that no apparent increase in galactoside yield for enzyme reactions in organic solvents. It was found that GOS formed in organic media would stay and not be hydrolyzed because the reaction was slowed down to a complete stop due to enzyme deactivation. Typically, the enzyme lost almost all its activity in only a few hours contact with the organic solvent. In general, conducting the reaction in solvents slowed the reaction rate and caused an accelerated enzyme deactivation (*69*). The lowered reaction rates and high enzyme costs would render this method uneconomical. Furthermore, the use of an organic solvent in GOS production may cause difficulty in product purification and raise severe environmental concerns.

Adding Galactose

Another method that could possibly increase GOS yield is the addition of galactose to a reaction system. Galactose inhibits the lactose hydrolysis reaction. Also, from thermodynamics point of view, a high galactose (product) concentration should favor the reverse hydrolysis reaction and cause an equilibrium shift towards condensation or transgalactosylation. This approach has been used in the synthesis of lactulose from galactose and glucose, although the product yield was low, less than 10% (71,72). Another drawback is that the reaction would be slow, due to the competitive inhibition effects of galactose on the enzyme.

The hypothesis that adding galactose would increase GOS yield was tested with *A. oryzae* β-galactosidase. The reactions were carried out with 40% lactose and 0 – 10% galactose. As expected, the reaction proceeded increasingly slower with increasing the amount of galactose in solution. Surprisingly, adding galactose did not increase the GOS yield, but instead decreased it (Figure 6). The higher the amount of added galactose, the lower the GOS production at all conversions. The GOS yield was reduced more at lower conversions, indicating that galactose actually inhibited the enzyme more in its transgalactosylation activity than its hydrolysis activity. Galactose inhibited the enzyme the most at the low lactose conversions where the transgalactosylation reaction dominated. This might be due to a limited supply of free glucose needed to start the transgalactosylation reaction.

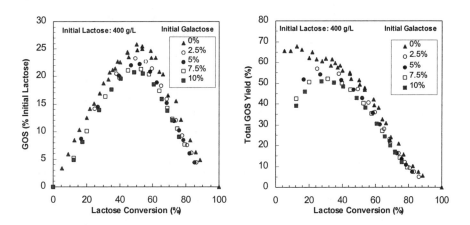

Figure 6. *Effects of galactose on GOS production. The experimental conditions were the same as those shown in Figure 1, except that the reaction medium initially also contained galactose in the amount as indicated.*

Removing Oligosaccharides

From thermodynamic and kinetic standpoints, removing oligosaccharides from the reaction medium should break the reaction equilibrium to favor more oligosaccharide formation (73). This hypothesis is supported by the observed higher GOS yield (up to ~70%) at low lactose conversions (and low GOS concentrations). Therefore, a GOS yield of greater than 70% is possible if GOS are removed continuously from the reaction medium.

However, the fact that the maximum GOS yield occurs at a low lactose conversion (~10%) presents a challenge to the design of a process with both high GOS yield and lactose conversion. Two possible processes are shown in Figure 7. The first is a single enzyme reactor (operating with a low lactose conversion) coupled to a continuous separation system to separate unreacted lactose and recycle it to the reactor inlet, allowing the process to have an overall high lactose conversion along with a high overall GOS yield. An efficient method to separate oligosaccharides from lactose, glucose, and galactose is needed for this process. The second process consists of multiple stages of reactor/separator. For this process, the overall process yield of GOS would be dependent on the extent of reaction (lactose conversion) at each stage and the total number of stages.

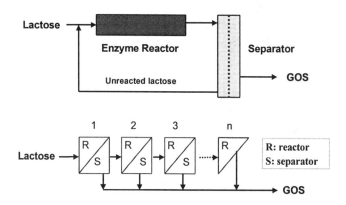

Figure 7. Continuous enzyme reaction and separation processes for GOS production from lactose.

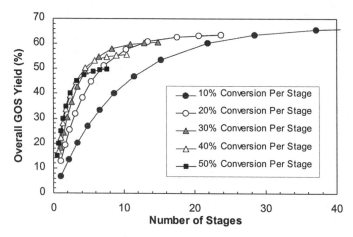

Figure 8. GOS yield as affected by the lactose conversion per stage and total number of stages in the continuous reaction-separation process.

As shown in Figure 8, a higher extent of reaction at each stage would require fewer stages to complete the reaction but the maximum GOS yield would be lower. For reactions with *A. oryzae* β-galactosidase, a GOS yield of 68% can be obtained in ~35 stages with 10% conversion per stage, whereas a process containing 10 stages at 20% lactose conversion per stage would give an overall GOS yield of 58%. These calculations were made by assuming perfect separation of lactose from other products. It must be noted that an inefficient separation process would greatly lower the overall process yield. Methods for GOS separation are discussed later in this article.

Immobilized β-Galactosidase

Enzyme immobilization is commonly employed during a continuous process; it typically increases enzyme life and allows for the enzyme, usually a large portion of the process costs, to be used repeatedly since it is retained in the reactor. This also allows for easier downstream processing of the products, and a greater control over the reaction.

Immobilization Methods

β-galactosidase has been immobilized on numerous supports, including glass beads, Celite, sintered alumina (*74*), chitosan beads (*35*), ion exchange resin Duolite ES-762 (*33,56,74*), Dowex MWA-1 (*33,74*), Amberlite XAD-2, XAD-7 and XAD-8 (*33*), Merkogel (*33,75*), and porous film (*76*). Enzyme also has been immobilized by entrapment in polymer gels such as agarose beads (*52*). The solid support for enzyme immobilization must be economical, stable, rigid, inert, and regenerable (*56*). The main difficulty in enzyme immobilization is allowing a large amount of the enzyme to be immobilized and retain all of its initial activity. Keeping the activity of the immobilized enzyme high and the actual retention of the enzyme on the support are also of concern, both long and short term.

The most popular enzyme supports are particulate, such as resins or beads. However, these supports may cause some problems when used in a continuous reactor, notably a high-pressure drop, ease of fouling, and possible plugging. One novel support gaining attention for the immobilization is porous fibrous matrix. The fibrous matrix, placed in a loose spiral in a plug-flow type reactor, has little diffusion limitations, good flow rates, and low pressure drop. However, there have been only a few studies on the immobilization of an enzyme to such fibrous supports. β-galactosidase was also successfully immobilized on hydrophobic cotton cloths by Sharma and Yamazaki (*77*), and Butler (*78*) bound 10 other enzymes to hydrophobic derivatives of cellulose materials including cotton string, cloth and filter paper; however, both of these studies used multi-step immobilization methods. Phosphatase and protease were bound covalently on silk fiber (*79*) and copoly(ethylene/acrylic acid) fibers (*80*), respectively. Shukla and Athalye (*81*) immobilized α-amylase on fibrous supports treated with glutaraldehyde, cyanogen bromide, UV radiation, and hydrogen peroxide, but the immobilization efficiency was very low. Cotton fibrous matrices are closely related to cellulose and chitosan, which has been successfully used to immobilize β-galactosidase. Sharp et al. (82) used N-(3-aminopropyl)-diethanolamine to attach enzyme to cellulose sheets (82) for immobilization. β-galactosidase was also successfully entrapped in cellulose fibers treated with titanium oxide in an acetone bath (83,84). For simple and novel immobilization of an enzyme to a fibrous support, the most successful outcome would probably be found via transition metal linking with $TiCl_4$, rivaling the retained enzyme activity in the difficult hydrophobic-cloth method (69).

Effects of Immobilization

Immobilization often changes an enzyme's configuration and rigidity, and ends up having a variety of effects on an enzyme's characteristics. It is often observed to shift an enzyme's optimal pH and temperature. In some cases where the enzyme can catalyze a multitude of reactions, immobilization can alter the distribution and production of the products. Thus, regioselectivity of oligosaccharide formation catalyzed by glycosidase can be affected by the structure of aglycons (85). Only a few authors have looked at β-galactosidase immobilization specifically for creating GOS as opposed to lactose hydrolysis. Enzyme immobilization has been found to affect oligosaccharide yield (52,75). Enzyme cross-linking with glutaraldehyde has been shown to increase GOS yield and affect the type of GOS produced (33,75). On the other hand, diffusion limitations in immobilized enzyme not only will reduce the reaction rate, but also may affect the product spectrum and reduce GOS formation. If the larger oligosaccharides diffuse slowly away from the enzyme, they may both inhibit transgalactosylation and be subjected to the hydrolysis reaction, lowering GOS yield. Mozaffar et al. (33) reported a gradual decrease in enzyme activity due to the accumulation of oligosaccharides inside the enzyme matrix (Merckogel). It is thus important to avoid diffusion limitations in immobilizing the enzyme in a matrix.

We have studied A. oryzae β-galactosidase immobilized on a phenolic fomaldehyde resin and on cotton cloth by TiO_2 treatment, and found that GOS production from the immobilized enzyme was almost unchanged as compared to free enzyme reaction (69).

Reactor Study

Several commercial processes using the immobilized lactase have been developed for lactose hydrolysis (21,22). However, only a few continuous reactors specifically for GOS production have been studied. Nakanishi et al. (74) and Mozaffar et al. (33) both studied continuous GOS production from B. circulans β-galactosidase; immobilization was on Duolite ES-762 and Merkogel, respectively. They both found that for the plug flow reactor (PFR), the enzyme activity gradually decreased over time, whereas in the continuous stirred tank reactor (CSTR) the immobilized enzyme seemed to be more stable initially but then activity quickly dropped when the glucose concentration reached a critical level. Both of them (*33,74*) reported that the deactivated enzyme in a PFR could be partially regenerated by passing buffer through the reactor for a day. Lopez-Levia and Guzman (*76*) developed a GOS production process using β–galactosidase immobilized on a thin (0.64 mm) porous film. However, this process had a low yield (33%) and is difficult to scale up, especially considering industrial factors such as high pressure (which could rupture a thin membrane), the large size of such a membrane, and fouling; continual replacement of such a film would be expensive.

More recently, Shin et al. (*35*) studied GOS production from lactose by *B. circulans* β-galactosidase immobilized in chitosan beads; they reported a GOS yield of 55% and reactor productivity of ~4.4 g/L/h. No loss of GOS production activity was observed for 15 days of continuous operation. Foda and Lopez-Leiva (*39*) studied continuous GOS production from whey lactose using a membrane reactor. The commercial enzyme from *K. lactis*, although not immobilized on a support, was kept in the reactor by the ultrafiltration (UF) membrane. Up to 31% GOS yield was obtained.

Oligosaccharides Separation

A high-efficient separation technique for recovering and purifying the various oligosaccharides from the reaction medium is needed. Different oligosaccharides may function differently and thus may have different pharmaceutical and medical applications. Oligosaccharide adsorption with activated charcoal and elution with alcohol solutions (5% and 60% ethanol) have been demonstrated and are currently used in the Japanese process (*86,87*). However, a much better separation method is needed in order to separate various oligosaccharides from lactose and monosaccharides present in mixture. An efficient separation technology will improve product yield, overall conversion of lactose as well as product qualities (such as purity). Analytical separation techniques such as HPLC, GC, TLC, and capillary electrophoresis (CE) have been used for oligosaccharide analyses (*88-90*). However, these laboratory techniques cannot easily be applied, if at all, to a large-scale process.

Recent developments in chromatographic separations allow efficient separation of chiral and many other isomeric compounds and oligosaccharides. The process-scale liquid chromatograph is now commonly used in the production of recombinant proteins in the biopharmaceutical industry. It is also used by the starch industry in the production of high fructose corn syrup, but it has not been used for oligosaccharide production. Aminex ion exchange resins, which are commonly used in HPLC for sugar analysis, can be used for this purpose.

An aqueous two-phase reaction system has been reported to selectively remove galactose from lactose, thus reducing galactose inhibition on lactose hydrolysis (*91*). The partition is based on the solubility difference between the two saccharides in the two aqueous phases. Since oligosaccharides have a lower solubility in a more hydrophilic solution than disaccharide and monosaccharides, it is possible to develop a two-phase reaction system that can partition the oligosaccharides of interest in one phase and the rest of saccharides and β–galactosidase in another phase, facilitating oligosaccharides synthesis. However, no work has been done in this area.

Nanofiltration (NF) is a relatively new membrane technology. Commercial NF membranes have a small pore size, usually about 2 nm, and molecular weight cutoffs (MWCO) in the range of 200 to 1000 Dalton. In NF, separation is guided by steric hindrance due to size, ionic charge, and hydrophobicity or solubility of the compound. NF has been successfully used in separating organic compounds with molecular weight of 200 to 500 Dalton, including organic acids (92), peptides and amino acids (93), chlorinated organic compounds (94), colors (95), and sugars (96,97) commonly found in food processing or industrial wastewater. Membrane processing, such as ultrafiltration, is widely used in the dairy industry (98). It should be feasible to use similar membrane separation technology (e.g., nanofiltration) to separate GOS from lactose and other saccharides on a large scale. Current research work shows that a typical NF run using a 300ml-capacity test cell can give 74.7% rejection for OS, 37.0% and 15.4% for lactose and glucose, respectively (76). In recovery of OS from steamed soybean wastewater with a spiral wound NF membrane, rejection for OS proves to be nearly 100% (99). Both indicate that NF could be an effective way to remove OS from the product stream and consequently could increase the overall yield of lactose into OS. By carefully selecting the NF membrane (pore size and charge) and solution conditions, one should be able to separate monosaccharides (MS) from disaccharides (DS) and OS or separate MS and DS from OS. But due to the small difference among MS, DS and OS, separation of OS from MS and DS or OS and DS from MS still remains a challenge, especially in a continuous operation mode (100).

Conclusion

The current status in the development of an enzymatic process to produce galacto-oligosaccharides (GOS) from lactose is reviewed. The enzyme β-galactosidase (lactase), commercially used to hydrolyze lactose in milk and whey, can be used to produce oligosaccharides containing 2 to 5 galactose units and one glucose unit from lactose. These GOS are natural constituents in mother's milk, have been found to efficiently and selectively accelerate the growth of *Bifidobacteria* in the lower intestine, and have many beneficial effects on human health. In this paper, we have also evaluated the feasibility of several novel approaches to enhance and to control GOS formation from lactose. It is concluded that an integrated immobilized enzyme reactor-separator process, which continuously removes GOS from the reaction media, would give the highest GOS yield (>65%) from lactose. Enzyme immobilization with appropriate support matrices is important to efficient GOS production. Porous fibrous matrices, including cotton towel, have advantages for use in this application due to their low pressure drop and high diffusion efficiency. Nanofiltration has a good potential for large-scale separation of GOS from lactose and monosaccharides (glucose and galactose), but further research and development in this area are needed. The GOS product can be used as a health-promoting food ingredient and dietary supplement from the surplus whey permeate and lactose currently produced in the dairy industry.

Acknowledgements

Financial support from USDA-CSREES (Grant No. 9801304) during this study and the *A. oryzae* lactase enzyme provided by Genencor International are acknowledged.

References

1. Mahoney, R.R. Galactosyl-oligosaccharide formation during lactose hydrolysis: a review. *Food Chem.* **1998**, 63, 147-154.
2. Crittenden, R.G.; Playne, M.J. Production, properties and applications of food-grade oligosaccharides. *Trends Food Sci. Technol.* **1996**, 7, 353-361.
3. Japan External Trade Organization. Oligosaccharides present opportunities for healthier food. *Tradescope* **1989**, 9, 20-22.
4. Tomomatsu, H. Health effects of oligosaccharides. *Food Technol.* **1994**, 48, 61-65.
5. Oku, T. Oligosaccharides with beneficial health effects: A Japanese perspective. *Nutr. Rev.* **1990**, 54, S59-S66.
6. Gibson, G.R.; Willis, C.L.; Van Loo, J. Non-digestible oligosaccharides and bifidobacteria - implications for health. *Int. Sugar J.* **1994**, 96, 381-387.
7. Chonan, O.; Matsumoto, K.; Watanuki, M. Effect of galactooligosaccharides on calcium absorption and preventing bone loss in ovariectomized rats, *Biosci. Biotech. Biochem.* **1995**, 59, 236-239.
8. Chonan, O.; Watanuki, M. Effect of galactooligosaccharides on calcium absorption in rats. *J. Nutr. Sci. Vitaminol.* **1995**, 41, 95-104.
9. Yazawa, K.; Imai, K.; Tamura, Z. Oligosaccharides and polysaccharides specifically utilized by bifidobacteria. *Chem. Pharm. Bull.* (Tokyo) **1978**, 26, 3306-3311.
10. Hoover, D.G. Bifidobacteria: activity and potential benefits. *Food Technol.* **1993**, 47, 120-124
11. Hughes, D.B.; Hoover, D.G. Bifidobacteria: their potential for use in American dairy products. *Food Technol.* **1991**, 45, 64-83.
12. Ishibashi, N.; Shimamura, S. Bifidobacteria: research and development in Japan. *Food Technol.* **1993**, 47, 126-135.
13. Tohyama, K.; Tanaka, R.; Kobayashi, Y.; Mutai, M. Relationship between the metabolic regulation of intestinal microflora by feeding *Bifidobacterium* and host hepatic function. *Bifidobacteria Microflora* **1982**, 1, 45-50.

14. Stahl, B.; Thurl, S.; Zeng, J.; Karas, M.; Hillenkamp, F.; Steup, M.; Sawatzki, G. Oligosaccharides from human milk as revealed by matrix-assisted laser desorption/ionization mass spectrometry. *Anal. Biochem.* **1994**, 223, 218-226.

15. Bouhnik, Y.; Flourie, B.; D'Agay-Abensour, L.; Pochart, P.; Gramet, G.; Durand, M.; Rambaud, J.-C. Administration of transgalacto-oligosaccharides increases fecal *Bifidobacteria* and modifies colonic fermentation metabolism in healthy humans. *J. Nutrition.* **1997**, 127, 444-448.

16. Rowland, I.R.; Tanaka, R. The effects of transgalactosylated oligosaccharides on gut flora metabolism in rats associated with a human faecal microflora. *J. Appl. Bacteriol.* **1993**, 74, 667-674.

17. Yanahira, S.; Morita, M.; Aoe, S.; Suguri, T.; Nakajima, I.; Deya, E. Effects of lactitol-oligosaccharides on the intestinal microflora in rats. *J. Nut. Sci. Vitaminol.* **1995**, 41, 83-94.

18. Tanaka, R.; Takayama, H.; Morotomi, M.; Kuroshima, T.; Ueyama, S.; Matsumoto, K.; Kuroda, A.; Mutai, M. Effects of administration of TOS and *Bifidobacterium breve* 4006 on the human fecal flora. *Bifidobacteria Microflora* **1983**, 2, 17-24.

19. Yun, J.W. Fructo-oligosaccharides – occurrence, preparation, and application. *Enzyme Microb. Technol.* **1996**, 19, 107-117.

20. Hidaka, H.; Hirayama, M.; Yamada, K Fructooligosaccharides enzymatic preparation and biofunctions. *J. Carbohydr. Chem.* **1991**, 10, 509-522.

21. Gekas, V.; Lopez-Leiva, M. Hydrolysis of lactose: a literature review. *Process Biochem.* **1985**, February, 2-12.

22. Axelsson, A.; Zacchi, G. Economic evaluation of the hydrolysis of lactose using immobilized β–galactosidase. *Appl. Biochem. Biotechnol.* **1990**, 24/25, 679-693.

23. Bakken, A.P.; Hill, C.G.; Amundson, C.H. Use of novel immobilized β–galactosidase reactor to hydrolyze the lactose constituent of skim milk. *Biotechnol. Bioeng.* **1990**, 36, 293-309.

24. Prenosil, J.E.; Stuker, E.; Bourne, J.R. Formation of oligosaccharides during enzymatic lactose: Part I: state of art. *Biotechnol. Bioeng.* **1987**, 30, 1019-1025.

25. Prenosil, J.E.; Stuker, E.; Bourne, J.R. Formation of oligosaccharides during enzymatic lactose: Part II: Experimental. *Biotechnol. Bioeng.* **1987**, 30, 1026-1031.

26. Zarate, S.; Lopez-Leiva, M.H. Oligosaccharide formation during enzymatic lactose hydrolysis: a literature review. *J. Food Protection.* **1990**, 53, 262-268

27. Yang, S.T.; Tang, I.C. Lactose hydrolysis and oligosaccharide formation catalyzed by beta-galactosidase. *Ann. N.Y. Acad. Sci.* **1988**, 542, 417-422.

28. Yang, S.T.; Okos, M.R. A new graphical method for determining parameters in Michaelis-Menten-type kinetics for enzymatic lactose hydrolysis. *Biotechnol. Bioeng.* **1989**, 34, 763-773.

29. Flaschel, E.; Raetz, E.; Renken, A. The kinetics of lactose hydrolysis for the β-galactosidase from Aspergillus niger, *Biotechnol. Bioeng.* **1982**, 24, 2499-2518.

30. Bakken, A.P.; Hill, C.G.; Amundson, C.H. Hydrolysis of lactose in skim milk by immobilized β–galactosidase (*Bacillus circulans*). *Biotechnol. Bioeng.* **1992**, 36, 293-309.

31. Kim, C.R.; Lee, S.R.; Lee, Y.K. Formation of galactooligosaccharides by the partially purified β-galactosidase from *Aspergillus niger* CAD 1. *Han'guk Ch'uksan Hakhoechi* **1990**, 32, 323-333 (in Korean).

32. Iwasaki, K.; Nakajima, M.; Nakao, S. Galacto-oligosaccharide production from lactose by an enzymatic batch reaction using β-galactosidase. *Process Biochem.* **1996**, 31, 69-76.

33. Mozaffar, Z.; Nakanishi, K.; Matsuno, R. Continuous production of galacto-oligosaccharides from lactose using immobilized β-galactosidase from *Bacillus circulans*. *Appl. Microbiol. Biotechnol.*, **1986**, 25, 224-8.

34. Rahim, K.A.A.; Lee, B.H. Specific inhibitory studies and oligosaccharide formation by β-galactosidase from psychrotrophic *Bacillus subtilis* KL88. *J. Dairy Sci.* **1991**, 74, 1773-1778.

35. Shin, H.-J.; Park, J.-M.; Yang, J.-W. Continuous production of galacto-oligosaccharides from lactose by *Bullera singularis* β-galactosidase immobilized in chitosan beads. *Process Biochem.* **1998**, 33, 787-792.

36. Stevenson, D.; Stanley, R.; Furneaux, R. Oligosaccharide and alkyl β–galactopyranoside synthesis from lactose with *Caldocellum saccharolyticum* β–glycosidase. *Enzyme Microbial Technol.* **1996**, 18, 544-549.

37. Ohtsuka, K.; Oki, S.; Ozawa, O.; Uchida, T. Isolation and cultural condition of galactooligosaccharide-producing yeast *Cryptococcus laurentii, Hakkokogaku* **1988**, 66, 225-233 (In Japanese).

38. Shukla, R.T. Beta-galactosidase technology: A solution to the lactose problem. *CRC Critical Rev. in Food Technol.* **1975**, 1, 325-356.

39. Foda, M.I.; Lopez-Leiva, M. Continuous production of oligosaccharides from whey using a membrane reactor. *Process Biochem.* **2000**, 35, 581-587.

40. Cruz, R.; D'Arcadia Cruz.V.; Belote, J.G.; De Oliveira Khenayfes, M.; Dorta, C.; Dos Santos Oliveira, L.H.; Ardiles, E.; Galli, A., Production of transgalactosylated oligosaccharides (TOS) by galactosyltransferase activity from *Penicillium simplicissimum*. *Brazil. Bioresour. Technol.* **1999**, 70(2), 165-171.

41. Boon, M.A.; van der Oost, J.; De Vos, W.M.; Janssen, A.E.M.; van 'T Riet, K. Synthesis of oligosaccharides catalyzed by thermostable β-glucosidase from *Pyrococcus furiosus*. *Appl. Biochem. Biotechnol.* **1999**, 75(2-3), 269-278.

42. Onishi, N.; Yokozeki, K. Gluco-oligosaccharide and galacto-oligosaccharide production by *Rhodotorula minuta* IFO879. *J. Ferment. Bioeng.* **1996**, 82(2), 124-127.

43. Roberts, H.R.; Pettinati, J.D. Concentration effects in the enzymatic conversion of lactose to oligosaccharides. *J. Agr. Food Chem.* **1957**, 5, 130-134.

44. Burvall, A.; sp, N.G.; Dahlqvist, A. Oligosaccharide formation during hydrolysis of lactose with *Saccharomyces lactis* lactase (Maxilact R.): Part 1. Food Chem. 4, 243-250.

45. Nakao, M.; Harada, M.; Kodama, Y.; Nakayama, T.; Shibano, Y.; Amachi, T. Purification and characterization of a thermostable β-galactosidase with high transgalactosylation activity from *Saccharopolyspora rectivirgula*. *Appl. Microbiol. Biotechnol.* **1994**, 40, 657-663.

152

46. Onishi, N.; Kira, I.; Yokozeki, K. Galacto-oligosaccharide production from lactose by *Sirobasidium magnum* CBS6803. *Lett. Appl. Microbiol.* **1996**, 23(4), 253-256.
47. Gorin, P.A.; Spencer, J.F.T.; Phaff, H.J. The structures of galactosyl-lactose and galactobiosyl-lactose produced from lactose by *Sporobolomyces singularis, Can. J. Chem.* **1964**, 42, 1341-1344.
48. Greenberg, N.A.; Mahoney, R.R. Formation of oligosaccharides by β-galactosidase from *Streptococcus thermophilus. Food Chem.* **1983**, 10, 195-204.
49. Onishi, N.; Tanaka, T. Galacto-oligosaccharide production using a recycling cell culture of *Sterigmatomyces elviae* CBS8119. *Lett. Appl. Microbiol.* **1998**, 26(2), 136-139.
50. Berger, J.L.; Lee, B.H.; Lacroix, C. Oligosaccharides synthesis by free and immobilized β–galactosidases from *Thermus aquaticus* YT-1. *Biotechnol. Lett.* **1995**, 17, 1077-1080.
51. Prakash, S.; Suyama, K.; Itoh, T.; Adachi, S. Oligosaccharide formation by *Trichoderma harzianum* in lactose containing medium. *Biotechnol. Lett. 1987*, 9(4), 249-52.
52. Berger, J. L.; Lee, B.H.; Lacroix, C. Immobilization of β–galactosidase from *Thermus aquaticus* YT-1 for oligosaccharide synthesis. *Biotechnol. Techniques* **1995**, 9, 601-606.
53. Onishi, N.; Tanaka, T. Purification and properties of a novel thermostable galacto-oligosaccharide-producing β–galactosidase from *Sterigmatomyces elviae* CBS8119. *Appl. Environ. Microbiol.* **1995**, 61, 4026-4030.
54. Onishi, N.; Yamashiro, A.; Yokozeki, K. Production of galacto-oligosaccharide from lactose by *Sterigmatomyces elviae* CBS8119. *Appl. Environ. Microbiol.* **1995**, 61, 4022-4025.
55. Genencor International, Inc. Product literature for "Fungal Lactase." Rochester, New York., 1996.
56. Yang, S.T.; Okos, M.R. Effects of temperature on lactose hydrolysis by immobilized beta-galactosidase in plug-flow reactor. *Biotechnol. Bioeng.* **1989**, 33, 873-885.
57. Monsan, P.; Paul, F. Enzymatic synthesis of oligosaccharides. *FEMS Microbiol. Rev.* **1995**, 16, 187-192.
58. Rastall, R.A.; Bucke, C. Enzymatic synthesis of oligosaccharides, *Biotechnol. Genetic Eng. Rev.* **1992**, 10, 253-281.
59. May, S. Biocatalysis in the 1990s: A perspective. *Enzyme Microb. Technol.* **1992**, 14, 80-84.
60. Zaks, A.; Klibanov, A.M. Enzymatic catalysis in nonaqueous solvents. *J. Biol. Chem.* **1988**, 263, 3194-3201.
61. Chaplin, M.F.; Bucke, C. *Enzyme technology;* Cambridge University Press: New York, NY. 1990; pp 220-225.
62. Gutman, A.L.; Shapira, M. Synthetic applications of enzymatic reactions in organic solvents. *Adv. Biochemical Eng./Biotechnol.* **1995**, 52, 88-128.
63. *Biocatalysis in Organic Media;* Laane, C.; Tramper, J.; Lilly, M.D., Eds.; Elsevier: Amsterdam, 1987.

64. Laroute, V.; Willemot, R. Effect of organic solvents on stability of two glycosidase and on glucoamylase-catalysed oligosaccharide synthesis. *Enzyme Microb. Technol.* **1992**, 14, 528-534.

65. Vulfson, E.N.; Patel, R.; Beecher, J.E.; Andrews, A.T.; Law, B.A.Glycosidase in organic solvents: I. Alkyl-beta-glucoside synthesis in a water-organic two-phase system. *Enzyme Microb. Technol.* **1990**, 12, 950-954.

66. Vulfson, E.N.; Patel, R.; Beecher, J.E.; Andrews, A.T.; Law, B.A. Glycosidase in organic solvents: II. Transgalactosylation catalysed by polyethylene glycol-modified beta-galactosidase. *Enzyme Microb. Technol.* **1990**, 12, 955-959.

67. Shin, H.-J.; Yang, J.-W. Galacto-oligosaccharide production by β–galactosidase in hydrophobic organic media. *Biotechnol. Lett.* **1994**, 16, 1157-1162.

68. Usui, T.; Kubota, S.; Ohi, H. A convenient synthesis of β–D-galactosyl disaccharide derivatives using the β–D-galactosidase from *Bacillus circulans*. *Carbohydr. Res.* **1993**, 244,:315-323.

69. Bednarcik, J.A. M.S. Thesis, The Ohio State University, Columbus, Ohio, 1998.

70. Stevenson, D.E.; Stanley, R.A.; Furneaux, R.H. Optimization of alkyl β–D–galactopyranoside synthesis from lactose using commercially available β–galactosidases. *Biotechnol. Bioeng.* **1993**, 42, 657-666.

71. Ajisaka, K.; Nishida, H.; Fujimoto, H. The synthesis of oligosaccharides by the reversed hydrolysis reaction of β–galactosidase at high substrate concentration and at high temperature. *Biotechnol. Lett.* **1987**, 9, 243-248.

72. Ajisaka, K.; Fujimoto, H. Regioselective synthesis of trisaccharides by use of a reversed hydrolysis of α- and β-D-galactosidase. *Carbohydr. Res.* **1989**, 185, 139-146.

73. Ekhart, P.F.; Timmermans, E. Techniques for the production of transgalactosylated oligosaccharides (TOS). *Bulletin IDF.* **1996**, 313, 59-64.

74. Nakanishi, K.; Matsuno, R.; Torii, K.; Yamamoto, K.; Kamikubo, T. Properties of immobilized β-D-galactosidase from *Bacillus circulans*. *Enzyme Microb. Technol.* **1983**, 5, 115-120.

75. Mozaffar, Z.; Nakanishi, K.; Matsuno, R. Effect of glutaraldehyde on oligosaccharide production by β-galactosidase from *Bacillus circulans*. *Appl. Microbiol. Biotechnol.* **1987**, 25, 426-429.

76. Lopez-Leiva, M.H.; Guzman, M. Formation of oligosaccharides during enzymic hydrolysis of milk whey permeates. *Process Biochem.* **1995**, 30, 757-762.

77. Sharma, S.; Yamazaki, H.. Preparation of hydrophobic cotton cloth. *Biotechnol. Lett.* **1984**, 6, 301-306.

78. Butler, L.G. Enzyme immobilization by adsorption on hydrophobic derivatives of cellulose and other hydrophilic materials. *Arch. Biochem. Biophys.* **1975**, 171, 645-650.

79. Asakura, T.; Kanetake, J.; Demura, M. Preparation and properties of covalently immobilized alkaline phosphatase on *Bombyx mori* silk fibroin fiber. *Polym.-Plast. Technol. Eng.* **1989**, 28, 453-469.

80. Emi, S.; Murase, Y. Protease immobilization onto copoly (ethylene/acrylic acid) fiber. *J. Appl. Polym. Sci.* **1990**, 41, 2753-2767.

81. Shukla, S.R.; Athalye, A.R. Immobilizing alpha-amylase onto fibrous supports. *Am. Dyestuff Rep.* **1995**, 84, 40-43.

82. Sharp, A.K.; Kay, G.; Lilly, M.D. The kinetics of β-galactosidase attached to porous cellulose sheets. *Biotechnol. Bioeng.* **1969**, 11, 363-380.
83. Kurokawa, Y.; Sano, T.; Ohta, H.; Nakagawa, Y. Immobilization of enzyme onto cellulose-titanium oxide composite fiber. *Biotechnol. Bioeng.* **1993**, 42, 394-397.
84. Ohmori, Y.; Kurokawa, Y. Preparation of fibre-entrapped enzyme using cellulose acetate-titanium-iso-propoxide composite as gel matrix. *J. Biotechnol.* **1994**, 33, 205-209.
85. Nilsson, K.G. A simple strategy for changing the regioselectivity of glycosidase-catalysed formation of disaccharides. *Carbohydr. Res.* **1987**, 167, 95-103.
86. Mutai, M.; Yamato, H.; Terashima, T.; Murayama, M.; Takahashi, T.; Tanaka, R.; Kuroda, A.; Ueyama, S.; Matsumoto, K. Composition for promoting growth of bifidobacteria. U.S. Patent 4,435,389, 1984.
87. Kaa, T.; Kobayashi, Y. Method for producing oligosaccharide. US Patent 4,957,860, 1990.
88. Betschart, H.F.; Prenosil, J.E. High-performance liquid chromatographic analysis of the products of enzymatic lactose hydrolysis. *J. Chromatography* **1984**, 299, 498-502
89. Nikolov, Z.L.; Meagher, M.M.; Reilly, P.J. Kinetics, equilibria, and modelling of the formation of oligosaccharides from D-glucose with *Aspergillus niger* glucoamylase I and II. *Biotechnol. Bioeng.* **1989**, 34, 694-704.
90. Oefner, P.; Chiesa, C.; Bonn, G.; Horvath, C. Developments in capillary electrophoresis of carbohydrates, *J. Capillary Electrophoresis*, **1994**, 1, 5-26.
91. Chen, J.P.; Wang, C.H. Lactose hydrolysis by beta-galactosidase in aqueous two-phase systems. *J. Fermentation Bioeng.* **1991**, 3, 168-175.
92. Jeantet, R.; Maubois, J.L.; Boyaval, P. Semicontinuous production of lactic acid in a bioreactor coupled with nanofiltration membranes. *Enzyme Microb. Technol.* **1996**, 19, 614-619.
93. Tsuru, T.; Shutou, T.; Nakao, S.; Kimura, S. Peptide and amino acid separation with nanofiltration membranes. *Sep. Sci. Technol.* **1994**, 29, 971-984.
94. Afonso, M.D.; Geraldes, V.; Rosa, M.J.; de Pinho, M.N. Nanofiltration removal of chlorinated organic compounds from alkaline bleaching effluents in a pulp and paper plant. *Wat. Res.* **1992**, 26, 1639-1643.
95. Fu, P.; Ruiz, H.; Lozier, J.; Thompson, K.; Spangenberg, C. Pilot study on ground water natural organics removal by low-pressure membranes. *Desalination* **1995**, 102, 47-56.
96. Aydogan, N.; Gurkan, T.; Yilmaz, L. Effect of operating parameters on the separation of sugars by nanofiltration. *Separation Sci. Technol.* **1998**, 33, 1767-1785.
97. Vellenga, E.; Tragardh, G. Nanofiltration of combined salt and sugar solutions: coupling between retentions. *Desalination* **1998**, 120, 211-220.
98. Rosenberg, M. Current and future applications fro membrane processes in the dairy industry. *Trends Food Sci. Technol.* **1995**, 6, 12-19.
99. Matsubara, Y.; Iwasaki, K.; Nakajima, M.; Nabetani, H.; Nakao, S. Recovery of oligosaccharides from steamed soybean waste water in Tofu processing by reverse osmosis and nanofiltration membranes. *Biosci. Biotech. Biochem.* **1996**, 60, 421-428.
100. Zhang, L. M.S. Thesis, The Ohio State University, Columbus, Ohio, 1998.

Chapter 10

Phyllosilicate Sol–Gel Immobilized Enzymes: Their Use as Packed Bed Column Bioreactors

A. F. Hsu, T. A. Foglia, K. Jones, and S. Shen

Eastern Regional Research Center, Agricultural Research Service, U.S. Department of Agriculture, 600 East Mermaid Lane, Wyndmoor, PA 19038

A novel procedure is described for the immobilization of the enzyme lipoxygenase (LOX) and the lipase from *Pseudomonas cepacia*. The enzymes were entrained within dispersed phyllosilicate clays that were cross-linked with silicate polymer formed by the controlled hydrolysis of tetramethyl orthosilicate (TMOS). The final activity of the phyllosilicate sol-gel immobilized enzymes was dependent upon the type of cross-linking reagent and catalyst used, and volume ratio of phyllosilicate clay to the free enzyme. Enzymes immobilized by this method were more stable, had higher activity than the free enzymes, and were reusable for at least five cycles without significant loss of activity. The feasibility of using these immobilized enzymes in packed-bed column bioreactors was demonstrated with immobilized LOX. This bioreactor was able to continuously produce hydroperoxy derivatives of polyunsaturated fatty acids. A simple computer simulation model was developed to determine the process kinetics. Both the experimental and model data indicated that immobilized LOX continuously catalyzes the oxygenation of linoleic acid.

Introduction

The immobilization of enzymes has been achieved by encapsulating them within a sol-gel matrix (*1*). Usually enzymes entrapped in sol-gel matrices retain much of their activity and have better stability than free enzymes. However, two disadvantages of sol-gel materials are their brittleness and narrow pore size (*2*). Efforts were made to improve the activity of immobilized enzymes by introducing additives, such as alginate (*3*) or polymers (2,4), into the sol-gel matrix. Recently, Reetz et al. (5) reported using different hydrophobic alkyl silanes as sol-gel precursors to entrap lipases with improved enzymatic activity. This latter method yielded immobilized lipases with eighty-fold activity in esterification reactions when compared to the free enzymes. More recently, we reported a novel procedure for the intercalative immobilization of lipoxygenase (LOX) within a phyllosilicate sol-gel matrix (6,7). This simple but effective procedure gave an immobilized lipoxygenase preparation with greatly improved activity and stability (6). The procedure uses alkylamines to occupy the charged sites of the phyllosilicate clay. This increases the hydrophobicity of the phyllosilicate clay, thus reducing the charge-charge interaction between the enzyme and the clay structure and enhancing enzyme activity. The application of this procedure for the immobilization of the lipase from *Pseudomonas cepacia* (PS-30) and LOX is reported here and the studies of the effect of reaction parameters on enzyme activity also are described.

Lipoxygenase is an enzyme that produces hydroperoxy fatty acid (HPOD) derivatives from polyunsaturated fatty acids, such as linoleic acid. HPOD derivatives are important intermediates for producing fungicides, pharmaceutical compounds, and other industrially important products (8,9,10, and 11). Since lipoxygenase is a labile enzyme, it is preferable to produce HPOD derivatives using immobilized lipoxygenase. Immobilization allows for enzyme reuse, enables continuous HPOD production under controlled conditions, and simplifies product isolation. Bioreactors can be designed on a laboratory scale in order to test the feasibility of the application of immobilized enzymes in industrial processes. Continuous reactors can be designed as either a column or packed bed reactor (12). This study was designed to determine whether the use of phyllosilicate sol-gel immobilized LOX in the packed bed column bioreactor is feasible for the continuous synthesis of peroxygenated fatty acids. A computer simulation model also is described that predicts process kinetics for a full-scale process.

Materials and Methods

Materials

Soybean lipoxygenase (type 1-B) and linoleic and lauric acids were obtained from Sigma (St. Louis, MO). *Pseudomonas cepacia* lipase (PS-30) was obtained from Amano Enzyme Co. (Troy, VA). Tetramethylorthosilicate (TMOS), trimethylammonium chloride (TMA), cetyltrimethylammonium chloride (HDTMA), and xylenol orange salt were purchased from Aldrich Chemical Co. (Milwaukee, WI). The phyllosilicate clay (montmorillonite, Wyoming, USA) was obtained from Source Clay Minerals Repository (Columbia, MO). All other reagents were analytical grades obtained from commercial sources.

Entrapment of Enzymes in Phyllosilicate Clay Cross-Linked with Sol-Gel

Phyllosilicate clay (type Swy-1) was saturated with sodium ions by three washes with NaCl solution (1M). Sodium ions in the clay were exchanged with alkyl ammonium ions by the addition of HDTMA or TMA (350 μL, 1M). The mixture was vortexed for a few seconds and then the clay was allowed to form a gel. Enzyme solution (lipoxygenase or lipase PS-30 in water, 1.2 mL containing 2-5 mg protein) was added followed by the addition of sodium fluoride (1M, 1 mL), TMOS (1.046 mL) and water (1.046 mL). The mixture was kept in the ice bath for 3 hours and then set at room temperature overnight to complete the polymerization process. The air-dried residue was washed with water to remove unretained enzyme and air-dried again. This residue was used for the phyllosilicate sol-gel immobilized enzyme studies.

Measurement of Lipoxygenase Activity

Lipoxygenase activity was measured spectrophotometrically using the xylenol orange method *(13)*. A 10-50 μL aliquot of the sample was added to 2 mL of 100 mM xylene orange reagent (composed of 250 M ammonium ferrous sulfate, 25 mM H_2SO_4 and 4 mM 2,6-di-t-butyl-4-methylphenol in methanol:water 90:10, v/v). The mixture was incubated at room temperature for 45 minutes and the absorbance at 560 nm was measured and compared to a freshly prepared cumene hydroperoxide standard curve. Protein content was measured using a modified Lowry procedure *(14)*.

Measurement of Lipase PS-30 Activity

Activity of immobilized lipase PS-30 was determined by following the esterification of lauric acid with 1-octanol. The immobilized enzyme (0.5 mg) was added to a mixture of 1-octanol (50 μL) and lauric acid (30 mg) in 3 mL of isooctane (water-saturated) in a 10 mL screw-capped Erlenmeyer flask. The reaction mixture was shaken for 18 h at 30°C. Aliquots were taken at selected time intervals and analyzed by gas chromatography (GC) to determine the extent of ester formation. Relative activity was calculated for all experiments by setting the activity of free PS-30 as 100%. Specific activities are defined as μmoles of octyl laurate formed per hour per mg of immobilized protein.

Bioreactor Design

The experimental apparatus used for the continuous peroxygenation of linoleic acid consisted of an enzyme reaction unit (packed with immobilized LOX), substrate reservoir and product detection unit (spectrophotometer and fraction collector), Figure 1. The substrate (linoleic acid; 0.5 mmole dissolved in 0.2M sodium borate buffer, 10 mM in sodium deoxycholate) in the reservoir (total volume: 300 mL) was circulated to the reaction unit by a high pressure pump (Waters, Milford, MA, USA) at a flow rate between 3.8 to 5.0 mL/min. Air or oxygen was continuously bubbled through the substrate reservoir to assure that the solution was oxygen-saturated. After the eluant circulated out from the reaction unit, a splitter divided the eluant into two parts: approximately 1/20 of the eluant flowed to the product detection unit with the reminder being circulated back to the substrate reservoir. Freshly prepared substrate solution was added from the storage unit to the substrate reservoir at a rate equal to the eluant flow rate to the product detection unit. Protein and HPOD content in the eluant were measured at 280 nm and 340 nm, respectively. HPOD content also was measured by the xylenol orange method described above.

Results and Discussion

Enzymatic Activity of Immobilized LOX

The data in Figure 2 show the relative activity of lipoxygenase (LOX) immobilized in sol-gel (TMOS alone, S), phyllosilicate sol-gel (CS) and trimethylammonium (TMA) saturated phyllosilicate sol-gel (CTS). For the TMA ion-saturated clay, the activity of immobilized LOX increased with the amount of TMOS used (treatment CST1 to CST3) for immobilization. The difference among the CST1, CST2 and CST3 preparations is the TMOS to clay ratio used. These values are 1, 5, and 10-volume percent respectively. As shown, the 10 volume percent TMOS preparation (CTS3) had the highest activity compared to the preparations obtained at the other volume ratios of TMOS to clay, clay with TMOS alone, and free LOX (Figure 2). No apparent increase in LOX activity was observed at higher TMOS to clay ratios. An opposite effect was observed for LOX immobilized in sodium ion-saturated clays (treatment CS1 and CS2). In this instance, increasing the TMOS to clay ratio from 1 to 5 volume percent (CS1 and CS2) caused a decrease in LOX activity. One possible explanation for this observation is that charge-charge interactions between the enzyme and clay inhibit the enzyme activity. Exchange of sodium ions by TMA ions reduces the cationic sites within the clay, thus reducing the net charge of the phyllosilicate and the charge-charge interactions with the enzyme.

Activity of Immobilized Lipase PS-30

Immobilization Conditions
As reported by Reetz (5), the hydrophobic environment of an immobilization matrix can influence the activity of the entrapped lipases. In that study, the preference of lipases for a more hydrophobic support was achieved by increasing the alkyl chain length of the alkyl silane used in sol-gel formation. In this report, we used alkylammonium ions, such as cetyltrimethylammonium (HDTMA) and TMA ions, to occupy the charge sites of the phyllosilicate clay, to increase the hydrophobicity of the clay. As shown in Figure 3, with TMA ion as the cationic exchange ligand, the activity of the immobilized lipase, measured by the esterification of lauric acid with 1-octanol, remained low but constant

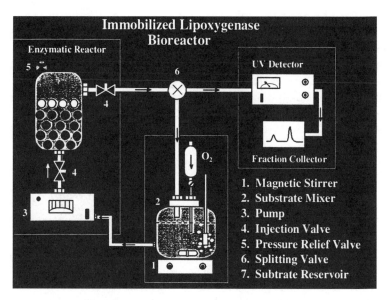

Figure 1. Lipoxygenase Bioreactor Design. 1: Magnetic stirrer. 2: Substrate mixer. 3: Pump. 4: Injection valve. 5: Pressure relief valve. 6: Splitting valve. 7: Substrate reservoir.

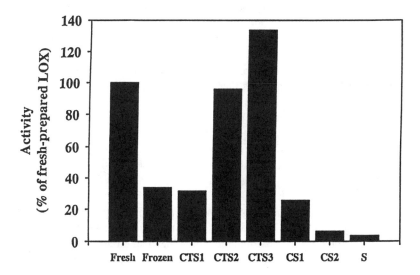

Figure 2. Activity of immobilized Lipoxygenase Preparations. The treatment codes are C: Clay, montmorillonite; T: TMA, trimethylammonium; S: TMOS, tetramethyl orthosilicate. Clay/TMOS ratios are CTS1 (CS1)=1, CTS2=5; CTS3=10. The immobilization of LOX in cation exchanged -phyllosilicate sol-gel matrix and the assay of LOX activity are described under "Materials and Methods."

over the range of 0-250 mM of TMA ions. In contrast, the use of HDTMA, a more hydrophobic cation exchange ligand, resulted in a significant increase in enzyme activity with optimum activity being obtained at 100 mM of HDTMA.

Although the data are not shown here, we also determined the optimal TMOS to clay ratios and sodium fluoride (NaF) TMOS polymerization catalyst concentration for immobilizing lipase PS-30 within a phyllosilicate sol-gel matrix. The results indicated that TMOS at 10% volume ratio (v/v of TMOS/total clay solution) and a NaF catalyst concentration of 100 mM gave the immobilized PS-30 preparation with the best activity.

Activity of Various Immobilized PS-30 Preparations

Figure 4 shows the time dependence of esterification of lauric acid with n-octanol using several immobilized lipase PS-30 preparations. The data show the sol-gel (TMOS alone) entrapped lipase PS-30 material (5) had low esterification activity. For PS-30 immobilized in a phyllosilicate sol-gel matrix using the procedure of Shen et al. (6) the lipase activity was moderate. When lipase PS-30 lipase was immobilized in a more hydrophobic sol-gel matrix [TMOS/PTMS (propyltrimethoxysilicate)] (5), its activity increased significantly and attained a maximum after 3.5 hours reaction. When lipase PS-30 was immobilized in a phyllosilicate sol-gel matrix using NaF catalyst as described herein, lipase activity was highest after 10 hours of incubation. To make valid comparisons, the protein content was constant for all experiments shown in Figure 4. As reported by Reetz et al. (5) when a lipase is immobilized within a TMOS/PTMS sol-gel its esterification activity is enhanced compared to the free lipase under the same conditions. The higher activity of TMOS/PTMS sol-gel immobilized lipases was attributed to the increased hydrophobic character of the sol-gel, which resulted in increased stability of the immobilized enzyme. In our method, we increased the hydrophobic character of the clay by exchanging sodium ions with TMA or HDTMA ions and then cross-linking the clay with the TMOS sol-gel. Because phyllosilicate clays are layered silicates with large surface area, these layered structures can be broken down to nanoscale building blocks that make them good matrices for entraining enzymes. As shown in Figure 4, the phyllosilicate sol-gel immobilized lipase PS-30 prepared in this study had the highest esterification activity of the immobilized lipases studied.

Reusability of Immobilized PS-30

The reusability of lipase PS-30 immobilized within a phyllosilicate sol-gel matrix in comparison to the free enzyme is shown in Figure 5. This was done by following the production of octyl laurate in the lipase-catalyzed esterification of lauric acid with 1-octanol. Ester formation was monitored by gas chromatography and the initial conversion was designated as 100% activity with subsequent reactions being compared to the initial reaction. After incubation, the immobilized and free PS-30 was recovered, washed with n-hexane, and dried under nitrogen. The next cycle of esterification was conducted with fresh reagents and the recovered enzyme. Data show that immobilized lipase P-30 could be used at least five times without loss of enzyme activity whereas the free enzyme lost most of its activity after the fifth cycle.

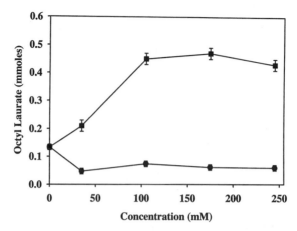

Figure 3. The effect of alkylammonium salts on immobilized PS-30 catalyzed esterification of lauric acid with 1-octanol. Alkylammonium reagents, HDTMA (■) and TMA (●) from 0-250 mM were added to the incubation mixture containing - TMOS and phyllosilicate suspension. The reaction mixture was then extracted and analyzed by gas chromatography (HP Innowax capillary column with temperature programming from 200 to 375 °C @ 5 °C per min). All analyses were performed in triplicate and experiments were conducted in duplicate.

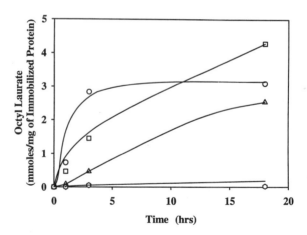

Figure 4. Kinetics of the esterification of lauric acid with 1-octanol catalyzed by immobilized PS-30. Reactions containing equal amounts of protein (0.5 mg) were incubated with lauric acid and 1-octanol as described under "Materials and Methods." Ester production was measured by gas chromatography. All experiments represent triplicate runs with standard error less than 5%. TMOS-immobilized PS-30 (●) TMOS/PTMS immobilized PS-30 (○); phyllosilicate sol-gel-immobilized PS-30 (Δ); phyllosilicate sol-gel/NaF-immobilized PS-30 (□).

Lipoxygenase Bioreactor for Continuous Oxygenation of Linoleic Acid

The experimental design of the bioreactor is shown in Figure 1. A computer program described by Folger and Brown (12) for determining the number of tanks in series to simulate a bioreactor was applied to this packed-bed bioreactor design. According to the mathematical calculation from the tracer experiment (data not shown), we calculated that the number of tanks of phyllosilicate sol-gel immobilized LOX needed to simulate this design was five. From mass balance equation (12), it was possible to calculate the reactant (linoleic acid) and product (hydroperoxide, HPOD) concentration in each reactor. The calculated results indicated that product formation in each tank increased as reaction time increased (data not shown).

The continuous oxygenation of linoleic acid by this phyllosilicate sol-gel immobilized LOX bioreactor produced HPOD in 0.2 M borate buffer containing 10 mM deoxycholate, as shown in Figure 6. An initial rapid dissociation of protein from the bioreactor also occurred but ceased after one hour. If fresh buffer was added into the system this protein loss could be reduced (data not shown). Figure 6 also shows the time course of HPOD production from this bioreactor. HPOD production rapidly increased for the first 2 hours, plateau then increased gradually thereafter. It can be seen from Figure 6 that the amount of HPOD from auto-oxidation (without linoleic acid) and background (without enzyme) were not significant enough to account for self-inactivation of the enzyme.

Conclusion

Enzymes, such as lipoxygenase and lipase PS-30, were successfully immobilized within phyllosilicate sol-gel matrices. The stability of these entrapped enzymes was greater than that of the free enzymes and the immobilized enzymes were reusable. The use of immobilized lipoxygenase in a packed-bed bioreactor demonstrated the continuous synthesis of peroxygenated fatty acids form linoleic acid. A computer-simulated model was developed to study the process kinetics of this bioreactor.

References

1. Avnir, D.; Braun, S.; Lev, O.; Ottolenghi, M. *Chem. Mater.* **1994**, *6*, 1605.
2. Heichal-Segal, O.; Rapppoport, S.; Braun, S. *Biotechnology* **1995**, *13*, 798.
3. Hsu, A-F.; Foglia, T. A.; Piazza, G. J. *Biotechnol. Lett.* **1997**, *19*, 71.
4. Shtelzer, S.; Rappoport, S.; Avnir, D.; Ottolenghi, M.; Braun, S. *Biotechnol. Appl. Biochem.* **1992**, *15*, 227.
5. Reetz, M.T.; Zonta, A.; Simpelkamp, J. *Biotechnol. Bioeng.* **1996**, *69*, 79.
6. Shen, S.; Hsu, A-F; Foglia, T. A.; Tu, S-I. *Appl. Biochem. Biotechnol.* **1998**, 69, 79.
7. Hsu, A-F, Shen, S.; Wu, E; Foglia, T. A. *Biotechnol. Appl. Biochem.* **1998**, *28*, 55.

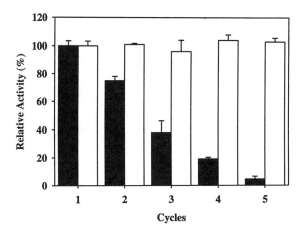

Figure 5. Recycling activity of Free PS-30 (open bars) and phyllosilicate sol-gel immobilized PS-30 (solid bars). Reactions were conducted with substrates (lauric acid and 1-octanol) in isooctane (water-saturated) for five cycles as described under "Materials and Methods." The first cycle of enzyme activity is expressed as 100%. Other activities are expressed relative to the first cycle.

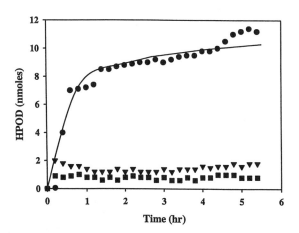

Figure 6. HPOD formation in phyllosilicate-LOX bioreactor. An aliquot from each fraction was assayed for HPOD concentration by the Xylenol Orange method. ●–● represents buffer containing linoleic acid; ▼–▼ is the circulating buffer without linoleic acid; and background (■–■) represents the determination of HPOD without the presence of immobilized enzyme in the bioreactor column.

8. Kato, T.; Maeda, Y.; Hirukawa, Y.; Namai, T.; Yoshikowa, Y. *Biosci. Biotechnol Biochem.* **1992**, *56*, 373.
9. Kato, T.; Yamaguchi, T.; Namai, Y.; Hirukawa, T. *Biosci. Biotechnol. Biochem.* **1993**, *57*, 283.
10. Vaugh, S.F.; Gardner, H.W. *J. Chem. Ecol.* **1993**, *19*, 2337.
11. Cuperus, F.P.; Kramer, G.F.H.; Derksen, J.T.P.; Bouwer, S.T. *Catalysis* **1995**, *25*, 441.
12. Fogler, H.S.; Brown, L.F. In *Analysis of Non-Ideal Reactors*; Folger, H.S., Ed.; Prentics Hall PTR, Englewood Cliffs, New Jersey, 1992.
13. Jiang, T.-Y.; Wollard, C.S.; Wolff, S.P. **1991**, *Lipids*, *26*, 853.
14. Bensadoun, A.; Weinstein, D. *Anal. Biochem.* **1976**, *70*, 214.

Chapter 11

Biocatalytic Preparation of D-Amino Acids

David P. Pantaleone, Paul P. Taylor, Tamara D. Dukes,
Kurt R. Vogler, Jennifer L. Ton, Daniel R. Beacom, Shelly L. McCarthy,
Bruce E. Meincke, L. Kirk Behrendt, Richard F. Senkpeil, and
Ian G. Fotheringham

NSC Technologies, A Division of Great Lakes Chemical Corporation,
601 East Kensington Road, Mt. Prospect, IL 60056

The use of D-amino acids in peptidomimetic pharmaceuticals has become common as a result of structure-based drug design. A novel enzyme system was used to prepare a number of D-amino acids starting from the inexpensive, L-isomer. This enzyme system is comprised of three enzymes: L-amino acid deaminase (L-AAD) from *Proteus myxofaciens,* D-amino acid aminotransferase (DAT) from *Bacillus sphaericus* and alanine racemase (AR) from *Salmonella typhimurium.* All enzymes were cloned into a single, *Escherichia coli* K12 strain and used as a whole-cell biocatalyst. Biotransformations were carried out to produce D-phenylalanine, D-tyrosine, D-tryptophan and D-leucine. This technology offers a competitive alternative for the production of D-amino acids compared to the classical resolution approaches currently practiced.

The rapid, worldwide increase in the development of single enantiomer drugs is placing increasing pressure upon fine chemical companies to develop new and more efficient synthetic routes to chiral intermediates. Unnatural amino acids are among the most prominent and diverse of the compound families occurring in the rational design of chiral drugs such as anti-cancer compounds and viral inhibitors. Accordingly, a variety of chemical and biological routes have been applied in the large scale manufacture of these compounds. In common with the commercial manufacture of natural amino acids, both chemo-enzymatic resolution approaches and direct single isomer syntheses have been undertaken in the production of unnatural amino acids and their derivatives (*1-3*).

166

Biocatalytic strategies are most successful when the inherent advantages of enzymes can be exploited to their fullest while minimizing potential drawbacks. High enantioselectivity, broad substrate specificity, high turnover and reusability are among the most significant benefits of enzymatic synthesis but can often be compromised by cofactor requirements and equilibrium constraints. Shown here are biosynthetic approaches which have been developed to prepare selected D-amino acids, which overcome traditional limitations and maximize the efficiency and versatility of our proprietary biocatalysts (*4, 5*).

Shown in Figure 1 is the general scheme used to convert L-amino acids into D-amino acids. L-Amino acids (L-phenylalanine, L-tyrosine, L-leucine) are converted to their respective α-keto acids by the action of the L-amino acid deaminase enzyme (L-AAD); a co-substrate for this reaction is molecular oxygen. The α-keto acid is then converted to the corresponding D-amino acid using D-alanine as the amino donor. This is catalyzed by the enzyme D-amino acid aminotransferase (DAT). The D-alanine is supplied as a racemate for economic reasons; the cloned alanine racemase gene provides a constant supply of D-alanine.

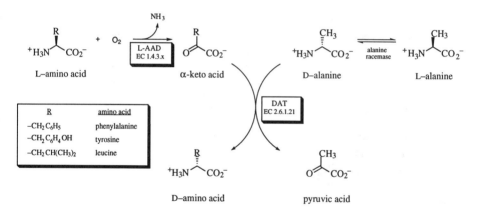

Figure 1. Reaction scheme for the whole cell biotransformation process to synthesize D-amino acids.

Materials and Methods

Chemicals and Solvents
All chemicals and solvents were purchased from commercial sources as reagent grade or better and were used without further purification.

Bacterial Strains and Plasmids
The host strain used in this study was an *E. coli* W3110 derivative carrying the following mutations: *dadAΔ aspCΔ*. The strain contains the following cloned genes

which have been deregulated: *dat* (D-aminotransferase) from *Bacillus sphaericus*, *alr* (alanine racemase) from *Salmonella typhimurium* and *aad* (amino acid deaminase) from *Proteus myxofaciens* (*6*). The host strain was named NS3353.

Media

Strains were routinely grown in Lennox L-broth or L-agar (1.5%) solid medium (Gibco/BRL) supplemented where appropriate with antibiotic. Fermentation medium was AG192 minimal medium (see Table I); antibiotics were used in the seed flasks, but not in the fermentation tanks. Following sterilization of the fermentors, sterile glucose was added aseptically to the fermentors to achieve a final concentration of 25 g/L.

Table I. AG192 Medium

Additive	Concentration per Liter
Magnesium sulfate•7H$_2$O	5.35 g
Ferric ammonium citrate	0.13 g
Postassium phosphate dibasic	4.60 g
Manganese sulfate (10 g/L stock)	2.3 mL
Potassium iodide/nickel sulfate (1.0 g/L stock)	0.74 mL
Trace metals solution[a]	0.66 mL
Antifoam DF204 (Mazur Mazu)	0.4 mL

[a]Contains 78.1 mg/L cobalt sulfate and sodium molybdate, 126.3 mg/L manganese sulfate and cupric sulfate and 609.4 mg/L zinc chloride.

Fermentation Conditions

Biolaffite fermentors (20L) were operated using a fed-batch process scheme under the following conditions: agitation, 500 rpm; temperature, 32°C; back pressure, 0.7 Bar; pH, controlled at 7.2 with gaseous ammonia; aeration, 1 vvm air; set volume, 10L; inoculation 1L. Following the depletion of the initial glucose, a feed stream of sterile glucose (70%, w/v) was supplied as required to maintain the dissolved oxygen at 15%. The fermentation was continued until the biomass achieved a wet cell weight (wcw) of 140 g/L.

Bioconversion Parameters

Standard conditions: 1L reaction volume (2L Biolaffite fermentor); 1400 rpm, 1.8 vvm air, L-phenylalanine (NSC Technologies), 2X molar equivalent D/L-ala (Penta), pH 8.0, 32°C, 100 g/L NS3353 cells, 24 hr reaction time. The L-phenylalanine concentration was greater than 0.2M and often exceeded solubility limits. The data are expressed as % conversion rather than in concentration due to the proprietary nature of this process, which is calculated as follows: (D-phenylalanine formed / L-phenylalanine utilized) x 100.

Variables tested: pH, temperature, cell loading, amino donor concentration, L-amino acid substrate and aeration rate.

Analytical

Chiral HPLC: Primary amine was derivatized using *o*-phthalaldehyde (Pierce) and BOC-cysteine (Novabiochem) (*7*) in borate buffer (pH 10.4) followed by separation on a C8, Columbus column (Phenomenex), 5μ, 100Å (250 x 4.6 mm). Mobile phase was 78% phosphate buffer (pH 6.2), 22% acetonitrile. Detection was by UV at 338 nm. Retention times are shown in Table II.

Table II. HPLC Retention Times in Minutes

Amino Acid	L-Isomer	D-Isomer
Alanine	5.15	6.25
Leucine	11.15	11.99
Phenylalanine	12.35	12.90
Tryptophan	9.87	10.62
Tyrosine	17.15	18.20

Chiral TLC: CHIRALPLATE Reversed Phase TLC (Alltech) - 50/50/200: methanol/water/acetonitrile. Detection with 0.2% (w/v) ninhydrin (Sigma) in ethanol followed by heating at 100°C for several minutes. Amino acid standards had the following R_f values: L-phenylalanine, 0.59; D-phenylalanine, 0.49; L-tyrosine, 0.66; D-tyrosine, 0.58.

Results

Enzyme Properties

L-Amino Acid Deaminase (L-AAD)

L-AAD (EC 1.4.3.x) was cloned from *Proteus myxofaciens* into *Escherichia coli*. This enzyme is membrane bound and functions with no added cofactors. The enzyme has a broad substrate specificity, (see Table II, preceding chapter by P. Taylor) and exhibits rapid reaction rates. The optimal pH range of the L-AAD reaction is 7.0 – 8.0. A strict O_2 requirement is observed and no H_2O_2 is formed during the reaction indicating that this enzyme is not a typical oxidase-type enzyme. This enzyme has also been referred to as an "oxidase" (*8-10*).

D-Amino Acid Aminotransferase (DAT)

DAT (EC 2.6.1.21) was cloned from *Bacillus sphaericus* into *Escherichia coli*. This enzyme also has a broad substrate specificity with an optimal pH range of 8.0 – 8.5. One of the best understood DATs is that from *Bacillus sp.* YM1, whose gene has been cloned and sequenced (*11*) and whose crystal structure has been solved (*12*). Other DAT genes have been cloned, sequenced and expressed in *E. coli* from *Staphylococcus haemolyticus* (*13*) and *Bacillus licheniformis* (*14*). Another DAT from *Bacillus brevis* and cloned into *E. coli* is currently being evaluated for its substrate specificity profile (N. Grinter, personal communication).

Alanine Racemase (AR)

AR (EC 5.1.1.1) was cloned from *Salmonella typhimurium* into *Escherichia coli*. This enzyme catalyzes pyridoxal 5′-phosphate-dependent conversion of L-alanine to D-alanine (*15*). Since the process utilizes racemic alanine, the presence of AR is necessary to provide a constant supply of D-isomer. It is a monomer with a molecular weight of about 40,000 and the pH optimum is 8.0 – 8.5.

Definition of Process Variables

The pH profile for the conversion of L-phenylalanine to D-phenylalanine is shown in Figure 2. The reaction scheme is that outlined in Figure 1. Bioconversion conditions are as follows: L-phenylalanine, 2X molar D/L-alanine, 100 g/L NS3353 cells, 32°C, 1.8 vvm air and 1400 rpm. pH was varied as indicated and maintained using NaOH. As shown in Figure 2, the pH optima for L-phenylalanine degradation is the same for D-phenylalanine synthesis, both being pH 8.0. The difference in

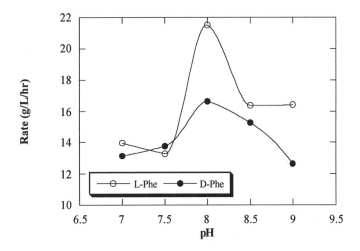

Figure 2. Initial rates of D-phenylalanine synthesis and L-phenylalanine consumption versus pH. Bioconversion conditions: L-phenylalanine, 2X molar D/L-alanine, 100 g/L NS3353 cells, 32°C, 1.8 vvm air and 1400 rpm. pH was adjusted and maintained with 10 N NaOH.

degradation and synthesis rates indicates that the DAT enzyme is rate limiting. Despite this rate difference, all subsequent bioconversions were carried out at pH 8.0.

The optimal cell loading was determined using differing levels of cells with the following process conditions: L-phenylalanine, 2X molar D/L-alanine, varying amounts of NS3353 cells, 32°C, 1.8 vvm air and 1400 rpm. As shown in Figure 3, the initial rates of conversion were similar for 50, 75 and 100 g/L cells; higher levels were tested (data not shown) resulting in poorer conversion to D-phenylalanine. Therefore, the optimal conversion was shown to be with 100 g/L cells.

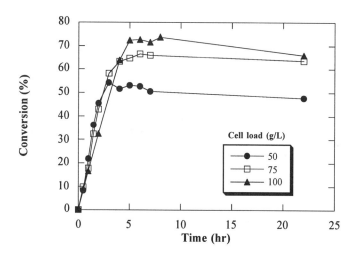

Figure 3. Bioconversion of L-phenylalanine to D-phenylalanine using different cell loadings of NS3353. Bioconversion conditions: L-phenylalanine, 2X molar D/L-alanine, pH 8.0, 32°C, 1.8 vvm air and 1400 rpm. Conversion (%) is based on the D-phenylalanine formed.

The oxygen dependence of the reaction was tested using different mixing rates and different aeration levels. The bioconversion conditions are as follows: L-phenylalanine, 2X molar D/L-alanine, 100 g/L NS3353 cells with varying aeration and rpm's as noted. Results are shown in Figure 4.

The biocatalyst stability was assessed after storage as concentrated cell slurries at 4°C for various times. At the indicated length of storage, a 1L bioconversion was carried out and the conversion to D-phenylalanine was monitored along with the L-phenylalanine remaining. Bioconversion conditions were as follows: L-phenylalanine, 2X molar D/L-alanine, 100 g/L NS3353 cells, 32°C, pH 8.0, 1.8 vvm and 1400 rpm. Results are shown in Figure 5, which indicated that after 13 days of storage, there is significant L-phenylalanine remaining even after 22 hours. This suggests that the L-AAD enzyme is inactivating and therefore, minimization of this storage time is required to ensure complete L-phenylalanine consumption.

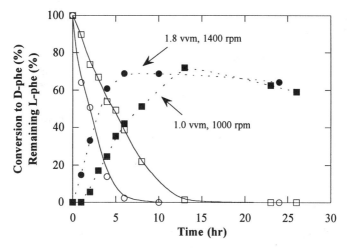

Figure 4. Oxygen dependence of L-phenylalanine consumption and D-phenylalanine production. Bioconversion conditions: L-phenylalanine, 2X molar D/L-alanine, pH 8.0, 32°C, 100 g/L NS3353 cells with varying aeration and rpm's. L-Phenylalanine is represented by open symbols and D-phenylalanine is represented by closed symbols. Aeration conditions are circles and squares for 1.8 vvm air, 1400 rpm and 1.0 vvm air, 1000 rpm, respectively.

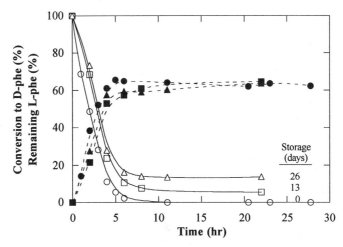

Figure 5. Bioconversion of L-phenylalanine to D-phenylalanine with cells of different age. Bioconversion conditions: L-phenylalanine, 2X molar D/L-alanine, pH 8.0, 32°C, 1.8 vvm air and 1400 rpm. Open symbols show the L-phenylalanine remaining (%) and closed symbols show the conversion to D-phenylalanine (%). Circles, squares and triangles represent 0, 13 and 26 days of storage, respectively.

Reaction Scale Up

Once the process variables of pH, aeration and cell loading were established, the reaction was scaled up to 20L. Reaction conditions were as follows: L-phenylalanine, 2X molar D/L-alanine, 100 g/L NS3353 cells, pH 8.0, 32°C, 1.0 vvm and 500 rpm. The dissolved oxygen (D.O.) was monitored throughout the reaction, which could be used to predict reaction status. Results shown in Figure 6 indicate that at nine hours, the dissolved oxygen spiked, all the L-phenylalanine was consumed and the D-phenylalanine conversion was nearly 80%. The enantiomeric excess of the product is >99%. The decrease in the D-phenylalanine conversion between nine and 22 hours is possibly due to sampling; product degradation has not been ruled out.

A further scale up to 200L was conducted to assess process variations and evaluate larger scale. Reaction conditions were as follows: L-phenylalanine, 2X molar D/L-alanine, 100 g/L NS3353 cells, pH 8.0, 32°C, 1.0 vvm and 500 rpm. D.O. (%) was continuously measured by a dissolved O_2 probe. As shown in Figure 7, all the L-phenylalanine is consumed by nine hours which corresponds to the D.O. spike. At the 9 hour time point, the conversion to D-phenylalanine is approximately 80% with an ee of >99%. It should be noted that the rpm's were reduced at 10.5 hours to minimize foaming.

Other Amino Acids

Other products, as indicated in Figure 1, can also be produced using this technology. As shown in Figure 8, D-tyrosine (D-tyr) and D-leucine (D-leu) can be produced along with D-phenylalanine (D-phe). Starting with the corresponding L-amino acid, NS3353 cells, at 100 g/L were used at pH 8.0 and 32°C at the 1L scale. D-tyrosine was prepared using 1X molar D/L-alanine while D-leucine and D-phenylalanine used 2X molar D/L-alanine. The aeration conditions to prepare D-tyrosine were 1.0 vvm and 1000 rpm, while those for D-leucine were 1.8 vvm and 1400 rpm. The enantiomeric excess for D-tyrosine, D-leucine and D-phenylalanine were 95.0, 97.5 and >99%, respectively.

These results show that preparation of other amino acids is feasible using this technology. Another amino acid, L-tryptophan was tested in this system, however, utilization of the L-isomer substrate appeared to level off after five hours with 25% of the starting substrate unreacted. Therefore, it appears that the L-AAD enzyme is inhibited by the substrate or the keto acid product and thus this system cannot be used to synthesize D-tryptophan under these conditions. Current studies are in progress to address this limitation and also to assess other potential L-amino acid substrates.

Discussion

The reaction conditions for the bioconversion of L-phenylalanine to D-phenylalanine have been explored and are discussed above. This complex technology which utilizes whole cells containing three cloned genes has also been demonstrated for two other amino acids, D-tyrosine and D-leucine (see Figure 8). It is clear from the results shown that there are a number of important factors which have a direct impact

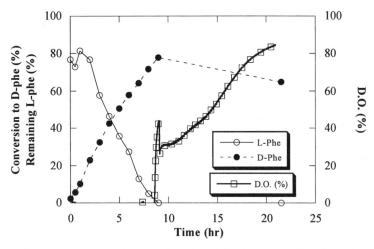

Figure 6. Bioconversion scale up of L-phenylalanine to D-phenylalanine to 20L. Bioconversion conditions: L-phenylalanine, 2X molar D/L-alanine, 100 g/L NS3353 cells, pH 8.0, 32°C, 1.0 vvm air and 500 rpm. Dissolved oxygen (D.O.) percentage was continuously monitored using a dissolved oxygen probe and the characteristic spike used as an in process test for reaction completeness.

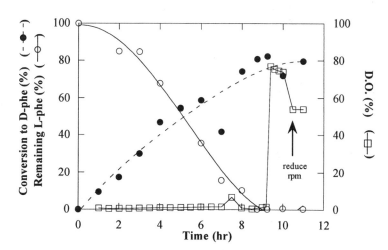

Figure 7. Scale up of L-phenylalanine to D-phenylalanine bioconversion to 200L. Bioconversion conditions: L-phenylalanine, 2X molar D/L-alanine, 100 g/L NS3353 cells, pH 8.0, 32°C, 1.0 vvm air and 500 rpm. . Dissolved oxygen (D.O.) percentage was continuously monitored using a dissolved oxygen probe and the characteristic spike used as an in process test for reaction completeness.

on the final conversion of product. A balance of activities is needed to ensure the best conversion to products and also to overcome any of the equilibrium issues which are known to exist for the transaminase enzymes (5).

Perhaps one of the most important parameters is the oxygen level in the system. This controls the rate of α-ketoacid product from the L-AAD reaction [phenylpyruvate (PPA) in the case of L-phenylalanine], which is supplied to the DAT enzyme for D-phenylalanine synthesis. It is important for two reasons: 1) high concentrations of PPA inhibit the L-AAD enzyme (data not shown); and 2) excess oxygen can cause oxidation of PPA to benzaldehyde (16). Therefore, it is imperative to run the reaction as near to 0% dissolved oxygen as possible to effectively control the overall coupled reaction by the rate of deamination. We have also observed that high aeration rates after product synthesis can lead to D-phenylalanine degradation over time (see Figure 6); the cause of this loss has not been determined.

Cell loading is crucial to the final enantiomeric excess (ee) of the product, which can clearly be seen in the cell loading study (see Figure 3); cell loading levels below 100 g/L yield low conversion and poor ee and those above 100 g/L do not increase conversion to product (data not shown). Biocatalyst stability is also an issue (see Figure 5) whereby storage of biocatalyst for more that a few days prior to use is not an option at this time due to the inability of the L-AAD enzyme to utilize all the L-phenylalanine. The reason for no significant increase in conversion using aged cells is not clear, but could possibly be due to an effect on the reaction equilibrium.

Finally, product solubility has a major effect on the total conversion of substrate to product. L-Tyrosine is very insoluble (~1.5 g/L in water) and as shown in Figure 8, an almost quantitative conversion to D-tyrosine can be realized. Preparation of D-phenylalanine and D-leucine (Figure 8), which are significantly more soluble than D-tyrosine, achieve conversions between 70 and 80%.

Overall, this coupled enzyme system is most effective for preparing D-amino acids from their respective L-isomers. This whole-cell bioprocess for D-phenylalanine was scaled up for pilot production at the 200L scale (see Figure 7). The conversion was nearly 80 % in nine hours with an ee of >99% and the characteristic D.O. spike at 9-10 hours can be used as an in-process measure of reaction completeness.

Current research is focusing on the L-AAD enzyme and what other amino acids might be utilized in this system. In addition, a more basic understanding of both L-AAD and DAT enzymes will allow better processes to be developed to prepare D-amino acids for use in chiral pharmaceuticals.

Literature Cited

1. *Chirality in Industry;* Collins, A. N.; Sheldrake, G. N.; Crosby, J., Eds.; John Wiley & Sons: New York, 1992.
2. *Chirality in Industry II;* Collins, A. N.; Sheldrake, G. N.; Crosby, J., Eds.; John Wiley & Sons: New York, 1997.
3. Wong, C.-H.; Whitesides, G. M. In *Enzymes in Synthetic Organic Chemistry;* Baldwin, J. E.; Magnus, P. D., Eds.; Tetrahedron Organic Chemistry Series; Elsevier Science Ltd.: Tarrytown, NY, 1994; Vol. 12.

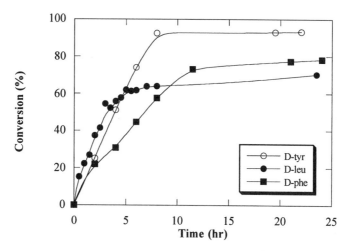

Figure 8. Comparison of D-tyrosine and D-leucine conversions to that of D-phenylalanine. Bioconversion conditions: L-amino acid, D/L-alanine (2X molar for phenylalanine and leucine; 1X molar for tyrosine), 100 g/L NS3353 cells, pH 8.0, 32°C 1.0 vvm air and various rpm's (1000 for tyrosine, 1400 for leucine and phenylalanine.

4. Fotheringham, I. G.; Pantaleone, D. P.; Taylor, P. P. *Chim. Oggi* **1997,** *Sep/Oct,* 33-37.
5. Taylor, P. P.; Pantaleone, D. P.; Senkpeil, R. F; Fotheringham, I. G. *TIBTECH* **1998,** *16,* 412-418.
6. Fotheringham, I. G.; Taylor, P. P.; Ton, J. L. U.S. Patent 5,728,555, 1998.
7. Brückner, H.; Wittner, R.; Godel, H. *J. Chromatogr.* **1989,** *476,* 73-82.
8. Pelmont, J.; Arlaud, G.; Rossat, A. M. *Biochimie* **1972,** *54,* 1359-1374.
9. Massad, G.; Zhao, H.; Mobley, H. L. T. *J. Bacteriol.* **1995,** *177,* 5878-5883.
10. Takahashi, E.; Furui, M.; Seko, H.; Shibatani, T. *Appl. Microbiol. Biotechnol.* **1997,** *47,* 173-179.
11. Tanizawa, K.; Asano, S.; Masu, Y.; Kuramitsu, S.; Kagamiyama, H.; Tanaka, H.; Soda, K. *J. Biol. Chem.* **1989,** *264,* 2450-2454.
12. Sugio, S.; Petsko, G. A.; Manning, J. M.; Soda, K.; Ringe, D. *Biochemistry* **1995,** *34,* 9661-9669.
13. Pucci, M. J.; Thanassi, J. A.; Ho, H.-T.; Falk, P. J.; Dougherty, T. J. *J. Bacteriol.* **1995,** *177,* 336-342.
14. Taylor, P. P.; Fotheringham, I. G. *Biochim. Biophys. Acta* **1997,** *1350,* 38-40.
15. Hayashi, H.; Wada, H.; Yoshimura, T.; Esaki, N.; Soda, K. *Ann. Rev. Biochem.* **1990,** *59,* 87-110.
16. Pitt, B. M. *Nature* **1962,** *196,* 272-273.

Applications: Pharmaceutical

Chapter 12

Application of Biocatalysis in Chemical Process Development: A Few Case Studies

Weiguo Liu[1]

NSC Technologies, A Unit of Monsanto Co., 601 East Kensington Road, Mt. Prospect, IL 60056

Two cases of applying biocatalysis for large scale production of chiral intermediates are discussed as part of the overall chemical process development stratagy for achieving practical and cost effective synthesis of the target biologically active compounds. A chemo-enzymatic synthesis of α,α-disubstituted α-amino acids was developed for the production of a key chiral intermediate in the synthesis of an anticholesterol drug candidate under clinical development. Another example involves the large scale preparation of optically active anti-viral nucleoside analogs through enzyme catalyzed kinetic resolutions. In both cases, the practical issues relating to large scale biocatalytic reaction are discussed in comparison with other alternative technologies such as classic resolution and asymmetric synthesis.

The use of enzymes and microorganisms in organic synthesis, especially in the production of chiral organic compounds, has grown significantly in recent years, and has been accepted as an effective and practical alternative for certain synthetic organic transformations. For an enzymatic resolution to be performed effectively on large scale, issues like solvent selection, volume through-put, product isolation, enzyme recycle, and the recycle of unwanted enantiomer have to be dealt with to achieve the acceptable economy. Very often, these evaluations are carried out in comparison with the results obtained using other alternative technologies such as classic resolution and asymmetric synthesis. Traditionally, research and development work on biocatalysis and biotransformations are carried out in a bioprocess laboratory setting since the

[1]Current address: Merck & Co., Inc., P.O. Box 2000, Rahway, NJ 07065.

isolation of new enzymes and handling of microorganisms and fermentation requires special sets of skills. With recently rapid advances in this field, a very large number of industrial enzymes have become commercially available, as well as being affordable in large quantities. This change has prompted many companies to set up biocatalysis groups within the chemical process development laboratories, so that all available technologies relating to the preparation of chiral organic compounds are directly accessible for a specific development project, and are coordinated based on the needs of the project. In this chapter, we will review a few cases of chemical development projects where one or more biocatalytic steps were applied to help achieve practical, efficient, and cost effective synthesis of the target chemical entities.

Synthesis of α,α-Disubstituted α-Amino Acids

Chiral α,α-disubstituted α-amino acids are important building blocks for the synthesis of pharmaceuticals and other biological agents (*1*). For the synthesis of an anticholesterol drug under development at the then Burroughs Wellcome Co. as 2164U90 (**1**), optically pure (*R*)-2-amino-2-ethylhexanoic acid (**2**) was needed as a critical building block (Scheme 1), and a campaign was launched for the development of a cost effective process for the large scale preparation of this chiral disubstituted amino acid and its analogs.

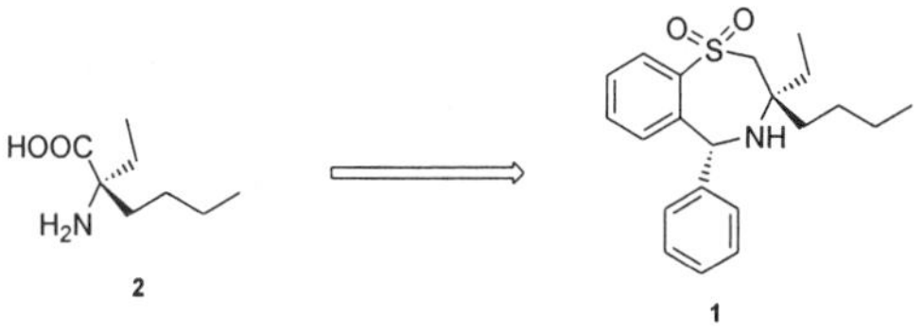

Scheme 1.

A variety of methods exists for the synthesis of optically active amino acids including asymmetric synthesis (*2-10*), and classic and enzymatic resolutions (*11-14*). However, most of these methods are not readily applicable to the preparation of α,α-disubstituted amino acids due to poor stereoselectivity and lower activity at the α-carbon. Attempts to resolve the racemic 2-amino-2-ethylhexanoic acid and its ester through classic resolution failed. Several approaches for the asymmetric synthesis of the amino acid were evaluated including alkylation of 2-aminobutyric acid using a camphor-based chiral auxiliary and chiral phase-transfer catalyst. A process based on Schöllkopf's asymmetric synthesis was developed as shown in Scheme 2 (*15*). Formation of piperazinone **3** through dimerization of methyl (*S*)-(+)-2-aminobutyrate (**4**) was followed by enolization and methylation to give (3*S*,6*S*)-2,5-dimethoxy-3,6-diethyl-3,6-dihydropyrazine (**5**) (Scheme 2). This dihydropyrazine intermediate is

180

unstable in air, and can be oxidized by oxygen to pyrazine **6** which has been isolated as a major impurity.

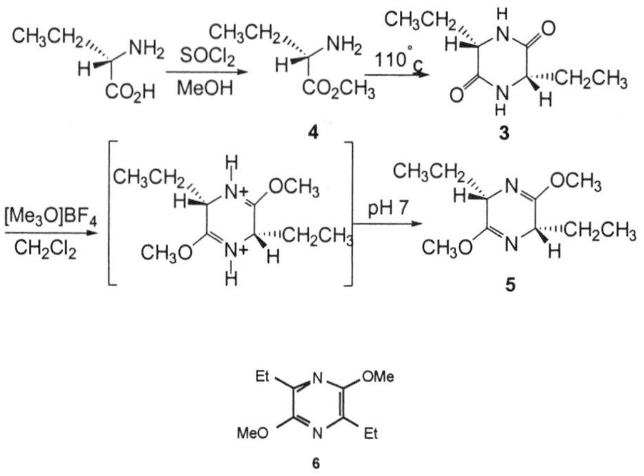

Scheme 2.

Operating carefully under nitrogen, compound **5** was used as a template for diastereoselective alkylation with a series of alkyl bromides to produce derivatives **7** which were then hydrolyzed by strong acid to afford the dialkylated (*R*)-amino acids **8** (~90% ee), together with partially racemized 2-aminobutyric acid which was separated from the product by ion-exchange chromatography (Scheme 3). This process provided gram quantities of the desired amino acids as analytical markers and test samples, but was neither practical nor economical for large scale production.

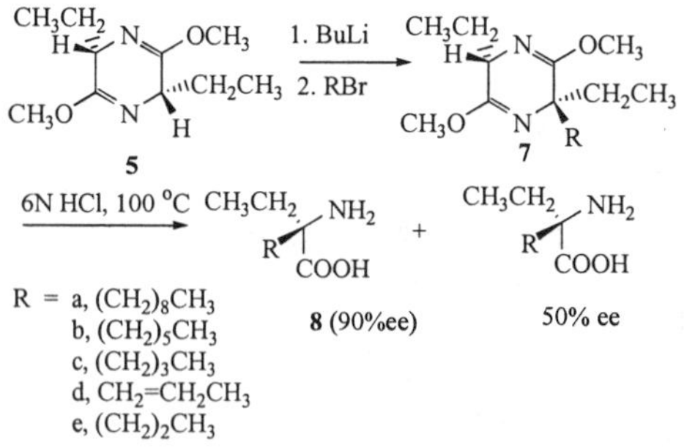

Scheme 3.

Facing difficulties with chemical methods, development work was aimed at finding a biocatalytic process to resolve the amino acids. Typical commercial enzymes reported for resolution of amino acids were tested. Whole cell systems containing hydantoinase were found to produce only α-monosubstituted amino acids (16-22); the acylase catalyzed resolution of N-acyl amino acids had extremely low rates (often zero) of catalysis toward α-dialkylated amino acids (23,24); and the nitrilase system obtained from Novo Nordisk showed no activity toward the corresponding 2-amino-2-ethylhexanoic amide (25,26). Finally, a large scale screening of hydrolytic enzymes for enantioselective hydrolysis of racemic amino esters was carried out. Racemic α,α-disubstituted α-amino esters were synthesized by standard chemistry through alkylation of the Schiff's base of the corresponding natural amino esters (Scheme 4), or through formation of hydantoins (27).

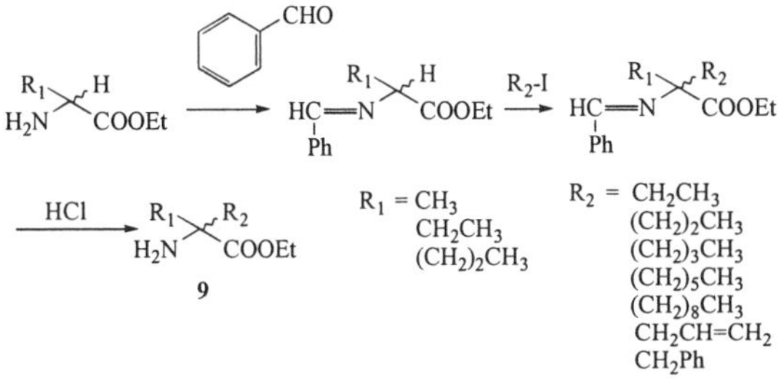

Scheme 4.

Initial enzyme screening was aimed at obtaining optically active 2-amino-2-ethylhexanoic acid (**8c**) or the corresponding amino alcohol. Enzymes reported for resolving α-H or α-Me analogs of amino acids failed to catalyze the corresponding reaction of this substrate (28), primarily due to the presence of the α-ethyl group that causes a critical increase in steric hindrance at the α-carbon. Out of 50 different enzymes and microorganisms screened, pig liver esterase and *Humicola lanuginosa* lipase (Lipase CE, Amano) were the only ones found to catalyze the hydrolysis of the substrate.

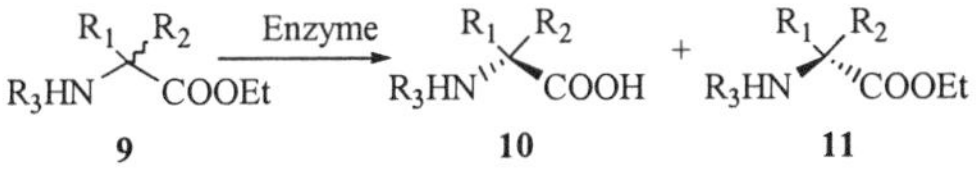

Scheme 5.

Both enzymes catalyze the hydrolysis of the amino ester (**9**) enantioselectively (Scheme 5). At about 60% substrate conversion, the enantiomeric excess of recovered ester (**11**) from both reactions exceeds 98%. In addition, the acid product (**10**) (96-98% ee) was obtained by carrying the hydrolysis of the ester to 40%. The rates of hydrolysis become significantly slower when conversion approaches 50%, allowing a wide window for kinetic control of the resolution process. Both enzymes function well in a concentrated water/substrate (oil) two-phase system while maintaining high enantioselectivity, making this system very attractive for industrial processes.

Although pig liver esterase catalyzed the hydrolysis of all amino esters tested in this work, it was only enantioselective towards esters (**9**, R_1 = $ArCH_2$, or n -Bu, R_2 = Et). This lack of correlation between enantioselectivity and substrate structure has been reported for many PLE catalyzed reactions (*29,30*). The lipase CE was obtained as a crude extract containing about 10% of total protein. The active enzyme catalyzing the amino ester hydrolysis was isolated from the mixture and partially purified. This new enzyme, *Humicola* amino esterase, is present as a minor protein component, has a molecular weight of about 35,000 daltons, and shows neither esterase activity toward *o*-nitrophenol butyrate nor lipase activity to olive oil. It is, however, highly effective in catalyzing the amino ester hydrolysis with very broad substrate specificity and high enantioselectivity. Various substrates including aliphatic, aromatic, and cyclic amino esters were resolved into optically active esters and acids (Table 1) with good E values (*31*). Aliphatic amino esters with alkyl or alkenyl side chains as long as 10 carbon atoms were well accepted by the enzyme. Substrates with longer chain length were limited by the solubility of the compounds rather than their binding with the enzyme. Resolutions were also extended to α,α-disubstituted amino esters in which the two alkyl groups differ in length by as little as a single carbon atom. The fact that the enzyme successfully catalyzes the resolution of straight chain aliphatic amino esters with two α-alkyl groups both larger than a methyl group is unique. These amino esters and their acids have been difficult to resolve by chemical and biochemical means due to the increased flexibility of the two large alkyl groups which become indistinguishable to most resolving agents. Unsubstituted amino esters underwent significant chemical hydrolysis under the experimental conditions, resulting in relatively lower E values.

When some amino esters were protected by *N*-acetylation, they become resistant to hydrolysis by the enzyme. Replacement of the α-amino group of the amino ester with a hydroxyl group also changed it from substrate to non-substrate. The enzyme showed high catalytic activity toward hydrolysis of phenylglycine ethyl ester, but was found incapable of catalyzing the hydrolysis of mandelic ethyl ester in which the amino group had been replaced with an α-hydroxyl group. The *N*-acetyl compounds and mandelic ester were not inhibitors, indicating that the free amino group is necessary for binding between the enzyme and substrates. It is most likely that the substrate is protonated at the amino group. The ammonium cation then binds to an anion on the enzyme's active site forming a strong ionic bonding to facilitate the catalysis (*Figure 1*).

Table 1: Enantioselective hydrolysis of α-amino esters catalyzed by *Humicola* amino esterase.

9, R_1 =	R_2 =	R_3 =	Conv. %	10 (%ee)	11 (%ee)	E value
CH$_3$(CH$_2$)$_8$-	Et	H	50	91 (R)[d]	--	58
CH$_3$(CH$_2$)$_5$-	Et	H	50	90 (R)[d]	99 (S)	58
CH$_3$(CH$_2$)$_5$-	Pr	H	37	92 (R)[c]	60 (S)	41
ArCH$_2$-	Et	H	50	85 (R)[a]	--	35
ArCH$_2$-	Me	H	74	32 (R)[a]	88 (S)	4.4
ArCH$_2$-	H	H	68	--	72 (S)[b]	4.0
Ar-	H	H	62	--	78 (S)[b]	6.3
CH$_3$(CH$_2$)$_3$-	Et	H	46	92 (R)[d]	--	58
			58	--	98 (S)[e]	
CH$_3$(CH$_2$)$_3$-	Et	OCOCH$_3$	NR			
CH$_3$CH=CH-	Et	H	42	94 (R)[d]	66 (S)	66
CH$_3$(CH$_2$)$_2$-	Et	H	65	26 (R)[d]	53 (S)	2.6
Me	Et	H	72	36 (S)[a]	100 (R)	20
H	Pr	H	40	72 (S)[b]	42 (R)	9.6
H	Et	H	55	53 (S)[b]	42 (R)	6.1
H	Et	OCOCH$_3$	NR			

NOTE: Absolute configurations were assigned by: [a] comparison of optical rotation with literature data; [b] comparison of HPLC retention time with authentic samples; [c] analogy to the results obtained from d; [d] comparison of HPLC retention time of its diastereomeric derivative with those obtained through Schollkopf's asymmetric synthesis; [e] X-ray crystallographic analysis.

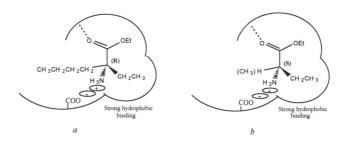

Figure 1. proposed substrate binding site for Humicola Amino Esterase
a. α,α-dialkyl amino esters with both side chains large than a methyl group
b, α-methyl or unsubstituted amino esters

This mechanism was further supported by the strong pH dependence of the enzymatic activity toward amino ester hydrolysis. Stability and relative activity of the amino esterase were measured over a range of pH's with ethyl 2-amino-2-ethylhexanoate (**9**, R_1 = Et, R_2 = *n*-Bu) as substrate and the data were compared as in Figure 1. The enzyme has an optimum pH of 7.5 and is most stable at pH 8. At low pH's, both the stability and activity of the enzyme suffered a gradual loss; this is consistent with the effect of gradual protein denaturation. At high pH's however, the enzyme remained relatively stable but its activity decreased sharply. The most significant drop of activity occurs when the medium's pH is greater than 9.6 that coincides with the substrate's pKa. Clearly when the substrate exists mostly in the form of a free base, its affinity with the enzyme dramatically diminishes at the active site where a charged ammonium cation is required for binding.

Initial scale up of the enzymatic resolution for production of kilogram quantities of (*R*)-2-amino-2-ethylhexanoic acid was performed in a batch process. The oil of ethyl 2-amino-2-ethylhexanoate was suspended in an equal volume of water containing the enzyme. The enantioselective hydrolysis of the ester proceeded at room temperature with titration of the produced acid by NaOH through a pH stat., Significantly, the reaction mixture was worked up by removal of the unreacted ester by hexane extraction and concentration of the aqueous layer to obtain the desired (*R*)-amino acid. The process has a high throughput and was easy to handle on a large scale. However, due to the nature of a batch process, the enzyme catalyst could not be effectively recovered, adding significantly to the cost of the product. In the further scale up to 100 kg quantity, the resolution process was performed using a membrane bioreactor module as shown in Figure 2. The enzyme was immobilized by entrapment into the interlayer of the hollowfiber membrane. Water and the substrate amino ester as a neat oil or hexane solution were circulated on each side of the membrane. The ester was hydrolyzed enantioselectively by the enzyme at the membrane interface, and the chiral acid product diffused to the aqueous phase to be recovered by simple concentration or ion exchange chromatography (*Figure 2*). Optical purity of (*R*)-2-amino-2-ethylhexanoic acid obtained from this process ranged from 98-99% ee (*32*).

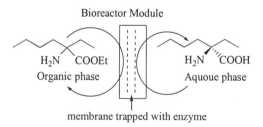

Figure 2. Membrane bioreactor for enzyme catalyzed enantioselective hydrolysis of racemic ethyl 2-amino-2-ethylhexanoate.

Synthesis of Optically Active Nucleoside Analogs

Nucleoside analogs with structural modifications either on the sugar or on the nucleobase or on both have been widely exploited as potential pharmaceutical agents for treatments of a variety of diseases. The ability of these compounds to interfere with DNA or RNA synthesis and thus tamper with the cell replication process makes them logical candidates of biological agents for treatment of cancer, viral infections and other genetic diseases. Several of these nucleoside analogs have been currently marketed as anti-viral agents for treatment of HIV and HBV related diseases. Traditional chemical synthesis of nucleosides involves the coupling of nucleobases with sugar derivatives, generating multiple stereoisomers which are often extremely difficult to separate and results in very low yields of the desired products. Applying biocatalysis at the appropriate steps in the synthetic sequence of the nucleosides can help separate the undesirable stereoisomers or simply eliminate their formation. Enzyme catalyzed enantioselective hydrolysis was used in the early chemical process development for production of the reverse trancriptase inhibitor drug 3TC, or Lamivudine, marketed by Glaxo Wellcome as Combivir, as well as FTC, or Coviracil, currently under late stage clinical development at Triangle Pharmaceuticals Inc. Both 3TC and FTC are oxathiolane nucleosides with 2-hydroxymethyl-5-oxo-1,3-oxathiolane as the sugar portion and cytosine or 5-fluorocytosine as the nucleobase (*33,34*). Both molecules contain two chiral centers and thus can exist in four different stereoisomeric forms. These drugs were originally developed as racemic mixtures of the β-anomers and were later switched to the optically pure forms of 2(R)-hydroxymethyl-5(S)-cytosinyl-1,3-oxathiolane duo to their favorable toxicology and pharmacokinetic profiles (*35-37*). There have been several independent routes leading to the synthesis of optically active 3TC or FTC, which involve either classic resolution, or enzymatic resolution, or asymmetric synthesis (*38-41*). Biocatalysis played significant roles in the early chemical development processes for production of both compounds in their optically pure forms, and thus allowed the supply of enough material for the initial preclinical and toxicological testing. Since both compounds were initially synthesized as racemic mixtures (*42*), a resolution process seems to be the most convenient way to obtain the desired optically active enantiomer in a short time frame. The enzyme catalyzed enantioselective reactions provided a safe, convenient, and practical way to effectively differentiate the two enantiomers resulting in their ultimate separation.

In the case of 3TC, the racemate was resolved through the enantioselective oxidation reaction catalyzed by cytidine deaminase (*43*). Cytidine deaminase and its counterpart adenosine deaminase belong to a class of nucleoside processing enzymes widely distributed in both mammals and microorganisms. Their biological roles are believed to be to scavenge exogenous and endogenous cytidine and adenosine for further metabolism through deamination of cytidine and adenosine to produce the corresponding uridine and inosine. Although these enzymes accept other unnatural nucleoside analogs with moderate modifications on both the nucleobase and the sugar portion, they observe strict conformational requirements. Only nucleosides with the same configuration or stereochemistry as cytidine or adenosine are accepted as substrates. Interestingly enough, both (-)3TC and (-)FTC, the enantiomers with the most desirable biological properties in the racemic mixture, have the opposite structural conformation relative to cytidine. Therefore, when the racemic mixture of 2'-deoxy-3'-thiocytidine was treated with cytidine deaminase, the undesired enantiomer with the same relative configuration as cytidine was deaminated to the corresponding uridine derivative while the desired enantiomer (-)3TC was untouched. The chemical structures of the deamination product and the unreacted starting material are different enough to allow separation of (-)3TC from the reaction mixture by conventional purification methods.

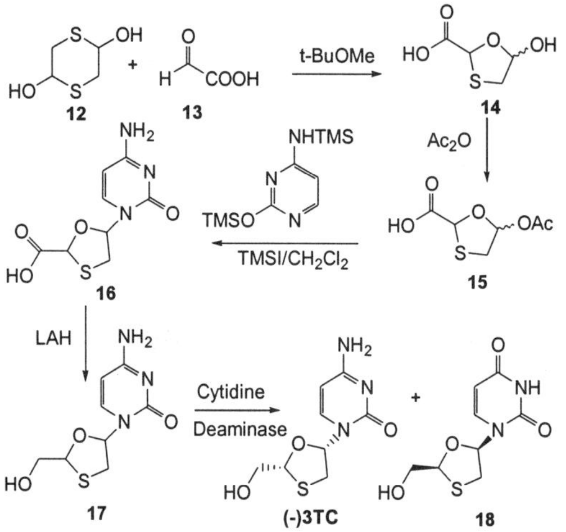

Scheme 6

As always, in a chemical process development setting, the success of any process is ultimately determined by how well it can be scaled up on a cost effective basis for production of large quantities of the product. For a biocatalytical process, the success often depends on the ready availability of the biocatalyst in large quantities. The cytidine deaminase used in 3TC production was cloned and over-expressed in E. coli, and produced in large quantities through conventional fermentation and purification. The isolated enzyme was then immobilized on to polymeric beads for repeated use and ease of product isolation. The general bioconversion procedures involve the suspension of the racemic mixture in water with pH adjusted to 7.0 by concentrated ammonium hydroxide. The immobilized enzyme is then added, and the reaction mixture stirred at 32 °C with addition of dilute acetic acid to maintain constant pH. The progress of the deamination reaction is monitored by the amount of acetic acid consumed and by taking samples periodically for chiral HPLC analysis. At about 52% conversion, essentially all the undesired enantiomer has been deaminated. The mixture is then filtered to recycle the enzyme. The filtrate is adjusted to pH 10.5, and passed through an anion ion-exchange column. Under these conditions, the unwanted deamination product is absorbed onto the column and unreacted (-)3TC passes through. The eluent was further desalted on a polystyrene resin column and the product washed out and recrystallized with acetone to obtain pure (-)3TC with greater than 99.8% enantiomeric excess (ee%). This process was scaled up for production of multi-kilogram quantities of the optically pure (-)3TC.

As for synthesis of the racemic starting material, there have been several alternative routes. The most efficient one starts with glycolic acid (13) and 1,4-dithiane-2,5-diol (12) which couples to form the oxathiolane ring system (14). The oxathiolane derivative is then acetylated and coupled with the protected cytosine to give the basic skeleton of 3TC (*Scheme 6*).

Since the chemical structures of 3TC and FTC are almost identical, their synthetic routes are generally interchangeable including the enzymatic resolution. Although the optically pure FTC, the 5-fluorosubstituted version of 3TC, can also be obtained through cytidine deaminase catalyzed resolution, its early production was achieved by a completely different synthetic route as shown in *Scheme 7 (36,42)*.

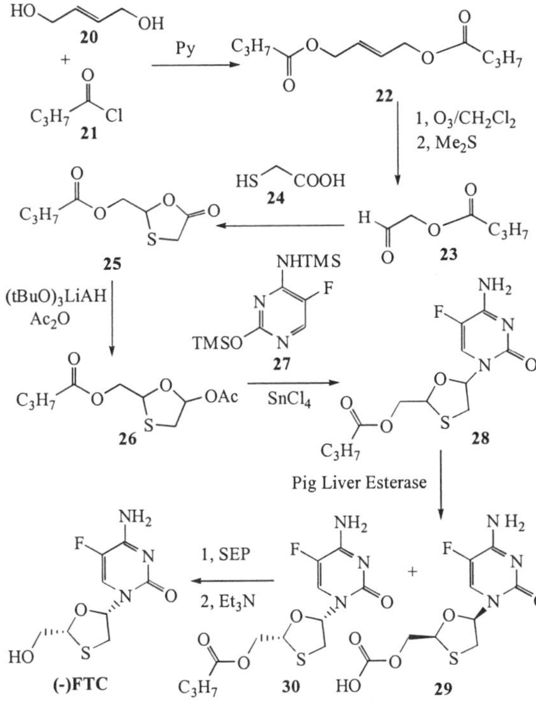

Scheme 7

Starting with 2-butene-1,4-diol (20), esterification of the diol with butyric chloride (21) followed by ozonolysis and reductive work up gave butyryl glycolaldehyde (23). The aldehyde was then condensed with mercaptoacetic acid (24) to form the lactone (25) which was reduced with lithium-tri-*t*-butoxyaluminum hydride followed by acetylation with acetic anhydride to produce the 2-butyroxymethyl-5-acetoxy-1,3-oxathiolane (26). This sulfur containing sugar analog was then coupled with silylated 5-fluorocytosine (27) in the presence of tin chloride as the Lewis acid to give the racemic FTC butyric ester (28). The racemic ester (28) was then subjected to pig liver esterase catalyzed enantioselective hydrolysis which selectively hydrolyzes the undesirable enantiomer and leaves the "unnatural" (-)-FTC ester (30) intact (*44*). The unreacted and optically enriched ester (30) is easily separated from the hydrolyzed product (29) by simple extraction into organic solvent, and the optically pure (-)-FTC final product was then obtained by chemical hydrolysis of the butyric ester (30).

Comparing to the cytidine deaminase catalyzed resolution, the pig liver esterase process has several advantages. Large quantities of both crude and purified pig liver esterase are readily available from several commercial sources, and the product isolation is fairly straight forward. The critical issue in the scale-up of this biocatalytic process is the reaction volume or the product through-put. The FTC butyrate starting material is almost insoluble in water, and can only be dissolved to a meaningful concentration by addition of water-miscible organic co-solvent. After a series of testing, acetonitrile was found to be the best organic solvent for the purpose. The solubility of the ester in water increases with the amount of acetonitrile added. However the activity or stability of the enzyme decreases with the increase of acetonitrile in the reaction mixture. When the concentration of acetonitrile in water is greater than 25% by volume, the enzyme's activity is reduced dramatically to unacceptable levels. Again, a series of reactions was conducted to determine the best solvent composition for maximum MeCN concentration and minimum reduction of enzyme activity. The final process uses 20% acetonitrile in 0.1 M potassium phosphate buffer of pH 8.0 as the solvent which dissolves up to 25 g/L of the ester starting material. To further increase the product throughput, the process was started with a suspension of 50 g/L of substrate. When the hydrolysis proceeds to about 50% conversion, the reaction mixture became a clear solution which was extracted with organic solvent to recover the unreacted nucleoside ester as optically active enantiomer of greater than 98% ee. This chiral ester was then hydrolyzed in water with triethylamine followed by crystallization to give (-)-FTC in greater than 99% ee. The process was also scaled up to produce multi-kilogram quantities of the product.

Summary

After years of research and development, biocatalysis and biotransformations have become an integral part of synthetic organic chemistry. They have found widespread applications in industry, especially in the production of chiral fine chemicals and pharmaceuticals. Although there are many issues remain to be solved on individual basis, the technology has generally been accepted as practical and economical solutions to many synthetic challenges in chemical process development. Rather than compete with organic chemical synthesis, biocatalysis complements to the existing chemical methodology and provides an extra tool to synthetic chemists. The key to the success of biocatalysis in chemical process development seems to depend on the clear understanding of the potential and limits of the technology, and thorough evaluation of the over all synthetic strategy so that one or more bio-steps can be applied where they make the most sense and bring the most value for the production of the target molecule.

Literature Cited

1. Coppola, G. M.; Schuster, H. F. *Asymmetric Synthesis: Construction of Chiral Molecules Using Amino Acids*; Wiley: New York, NY, 1987.
2. Williams, R. M.; Im, M.-N. *J. Am. Chem. Soc.* **1991**, *113*, 9276.
3. Schollkopf, U. *Pure Appl. Chem.* **1983**, *55*, 1799.
4. Schollkopf, U.; Busse, U.; Lonsky, R.; Hinrichs, R. *Liebigs Ann. Chem.* **1986**, 2150.
5. Seebach, D.; Boes, M.; Naef, R.; Schweizer, W. B. *J. Am. Chem. Soc.* **1983**, *105*, 5390.
6. Seebach, D.; Aebi, J. D.; Naef, R.; Weber, T. *Helv. Chim. Acta* **1985**, *65*, 144.
7. O'Donnell, M. J.; Bennett, W. D.; Bruder, W. A.; Jacobson, W. N.; Knuth, K.; LeClef, B.; Polt, R. L.; Bordwell, F. G.; Mrozack, S. R.; Cripe, T. A. *J. Am. Chem. Soc.* **1988**, *110*, 8520.
8. Kolb, M.; Barth, J. *Angew. Chem.* **1980**, *92*, 753.
9. Goerg, G. I.; Guan, X.; Kant, J. *Tetrahedron Lett.* **1988**, *29*, 403.
10. Obrecht, D.; Spiegler, C.; Schönholzer, P.; Müller, K.; Heimgartner, H.; Stierli, F. *Helv. Chim. Acta* **1992**, *75*, 1666.
11. Bosch, R.; Brückner, H.; Jung, G.; Winter, W. *Tetrahedron* **1982**, *38*, 3579.
12. Anantharamaiah, G. M.; Roeske, R. W. *Tetrahedron Lett.* **1982**, *23*, 3335.
13. Berger, A.; Smolarsky, M.; Kurn, N.; Bosshard, H. R. *J. Org. Chem.* **1973**, *38*, 457.
14. Kamphuis, J.; Boesten, W. H. J.; Broxterman, Q. B.; Hermes, H. F. M.; Meijer, E. M.; Schoemaker, H. E. in *Adv. Biochem. Engineering Biotechnol.;* Flechter, A. Ed.; Springer-Verlag: Berlin, 1990; *Vol. 42.*
15. Schollkopf, U.; Hartwig, W.; Groth, U.; Westphalen, K.-O. *Liebigs Ann. Chem.* **1981**, 696.
16. Gross, C.; Syldatk, C.; Mackowiak, V.; Wagner, F. J. *Biotechnology* **1990**, *14*, 363.
17. Chevalier, P.; Roy, D.; Morin, A. *Appl. Microbiol. Biotechnol.* **1989**, *30*, 482.
18. West, T. P. *Arch. Microbiol.* **1991**, *156*, 513.
19. Shimizu, S.; Shimada, H.; Takahashi, S.; Ohashi, T.; Tani, Y.; Yamada, H. *Agric. Biol. Chem.* **1980**, 44, 2233.
20. Syldatk, C.; Wagner, F. *Food Biotechnol.* **1990**, *4*, 87.

21. Runser, S.; Chinski, N.; Ohleyer, E. *Appl. Microbiol. Biotechnol.* **1990,** *33,* 382.
22. Nishida, Y.; Nakamicho, K.; Nabe, K.; Tosa, T. *Enzyme Micro. Technol.* **1987,** *9,* 721.
23. Chenault, H. K.; Dahmer, J.; Whitesides, G. M. *J. Am. Chem. Soc.* **1989,** *111,* 6354.
24. Keller, J. W.; Hamilton, B. J. *Tetrahedron Lett.* **1986,** *27,* 1249.
25. Kruizinga, W. H.; Bolster, J.; Kellogg, R. M.; Kamphuis, J.; Boesten, W. H. J.; Meijer, E. M.; Schoemaker, H. E. *J. Org. Chem.* **1988,** *53,* 1826.
26. Schoemaker, H. E.; Boesten, W. H. J.; Kaptein, B.; Hermes, H. F. M.; Sonke, T.; Broxterman, Q. B.; van den Tweel, W. J. J.; Kamphuis, J. *Pure Appl. Chem.* **1992,** *64,* 1171.
27. Ware, E. *Chem. Rev.* **1950,** *46,* 403.
28. Yee, C.; Blythe, T. A.; McNabb, T. J.; Walts, A. E. *J. Org. Chem.* **1992,** *57,* 3525.
29. Toone, E. J.; Jones, J. B. *Tetrahedron: Asymmetry* **1991,** *2,* 201.
30. Lam, L. K. P.; Brown, C. M.; Jeso, B. D.; Lym, L.; Toone, E. J.; Jones, J. B. *J. Am. Chem. Soc.* **1988,** *110,* 4409.
31. Chen, C. S.; Fujimoto, Y.; Girdaukas, G.; Sih, C. J. *J. Am. Chem. Soc.* **1982,** *104,* 7294.
32. Liu, W.; Ray, P.; Beneza, S. A. *J. Chem. Soc., Perkin Trans. 1* **1995,** 553.
33. Coates, J. A. V.; Mutton, I. M.; Penn, C. R.; Storer, R.; Williamson, C. PCT International Patent Application WO 9117159 A1 19911114.
34. Liotta, D. C.; Schinazi, R. F.; Choi, W. B. U. S. Patent 5,210,085, 1997.
35. Rogan, M. M.; Drake, C.; Goodall, D. M.; Altria, K. D. *Anal. Biochem.* **1993,** 208, 343.
36. Liotta, D. C.; Schinazi, R. F.; Choi, W. B. PCT International Patent Application WO 9214743 A2 920903.
37. Furman, P. A.; Davis, M.; Liotta, D. C.; Paff, M.; Frick, L. W.; Nelson, D. J.; Dornsife, R. E.; Wurster, J. A.; Wilson, L. J. *Antimicrob. Agents Chemother.* **1992,** 36, 2686.
38. Goodyear, M. D.; Dwyer, P. O.; Hill, M. L.; Whitehead, A. J.; Hornby, R. Hallett, P. PCT International Patent Application No. WO 95-EP1503 950421.
39. Jin, H.; Siddiqui, M. A.; Evans, C. A.; Tse, H. L. A.; Mansour, T. S.; Goodyear, M. D.; Ravenscroft, P. Beels, C. D. *J. Org. Chem.* **1995,** 60, 2621.
40. Chu, C. K.; Beach, J. W.; Jeong, L. S. PCT International Patent Application WO 9210496 A1 920625
41. Beach, J. W.; Jeong, L. S.; Alves, A. J.; Pohl, D.; Kim, H. O.; Chang, C. N.; Doong, S. L.; Schinazi, R. F.; Cheng, Y. C.; Chu, C. K. *J. Org. Chem.* **1992,** 57, 2217.
42. Liotta, D. C.; Schinazi, R. F.; Choi, W. B. U. S. Patent Application 91-659760.
43. Mahmoudian, M.; Baines, B. S.; Drake, C. S.; Hale, R. S.; Jones, P.; Piercey, J. E.; Montgomery, D. S.; Purvis, I. J.; Storer, R.; Dawson, M. J.; Lawrence, G. C. *Enzyme Microb. Technol.,* **1993,** 15, 749.
44. Hoong, L. K.; Strange, L. E.; Liotta, D. C.; Koszalka, G. W.; Burns, C. L.; Schinazi, R. F. *J. Org. Chem.* **1992,** 57, 5563

Chapter 13

Method for the Stereoselective Production of Chiral Vicinal Aminoalcohols: Establishing Two Chiral Centers by Diastereoselective Reduction

J. David Rozzell

BioCatalytics, Inc., 39 Congress Street, Suite 303, Pasadena, CA 91105 (telephone: 818–841–0072; fax: 818–841–0011; email: davidrozzell@biocatalytics.com)

Pharmaceutically-active compounds products from a number of different therapeutic categories contain chiral vicinal aminoalcohol substructures. As a result, chiral vicinal aminoalcohols represent a class of key intermediates for the drug industry. Gaining control over the stereochemistry at chiral centers bearing both the amino and hydroxyl groups is crucial to the successful synthesis of these compounds. The key step in the synthetic method is the reduction of a 2-substituted-ß-keotester catalyzed by an alcohol dehydrogenase. This reaction creates two chiral centers simultaneously. The ability to recycle the nicotinamide cofactor is also important for ultimate success of this synthetic method, and alternatives for efficiently accomplishing cofactor recycling are described.

Chiral vicinal aminoalcohols are key building blocks for the production of a number of pharmaceutical and agricultural products. Important compounds that may be produced from chiral vicinal aminoalcohol precursors span a range of therapeutic categories and include pseudoephedrine, epinephrine, norepinephrine, taxol, amistatin, betastatin, and the HIV-RT inhibitors indinavir (2), saquanivir (3), ritonavir (4), and nelfinavir (5). The structures of key aminoalcohol intermediates that may be used for the production of some key pahrmaceutical products are shown in Figure 1.

192

Figure 1: HIV Protease Inhibitors Based on Chiral Aminoalcohol Precursors

Protease Inhibitor	Aminoalcohol Intermediate
Indinavir Merck	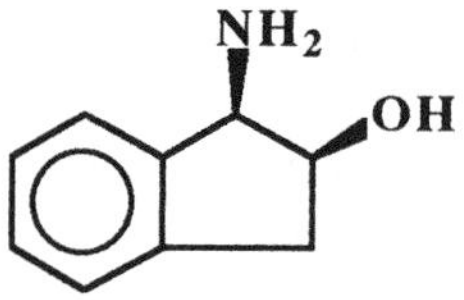
Saquanivir Roche	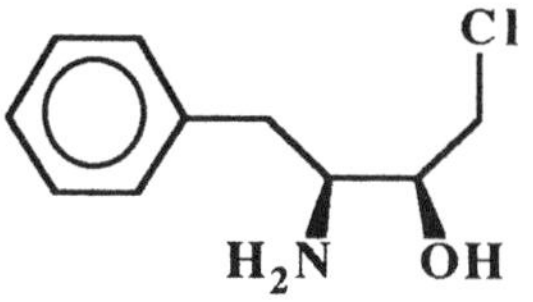
Ritonavir Abbott	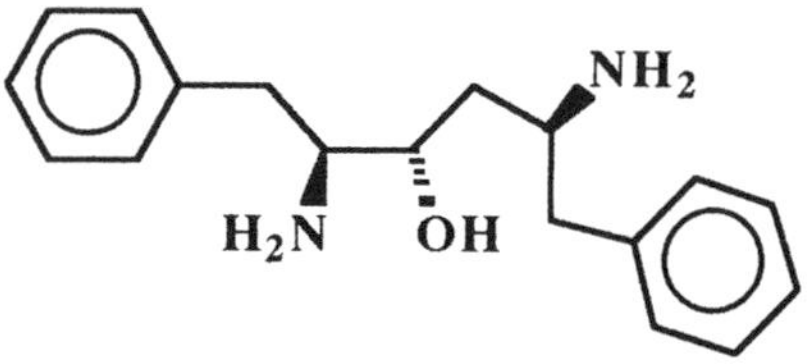
Nelfinavir Agouron	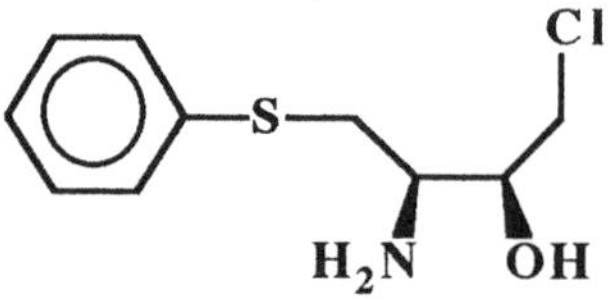

We describe here the key step in a process that may be employed to produce a broad range of chiral vicinal aminoalcohols, both cyclic and acyclic. A particularly notable feature of this process is the ability to control the absolute configuration at chiral centers bearing both the amino and alcohol functionality. A single diastereomer of the four possible stereoisomers can be produced in high stereochemical purity. This process has been granted U.S. Patent 5, 834,261, issued in November 1998. (1) A further patent relating to the key intermediates in the production of chiral vicinal aminoalcohols issued in August 1999. (2)

The process for producing chiral vicinal aminoalcohols relies on ß-ketoesters as starting materials. These compounds are generally readily available, with the simplest example being acetoacetic ester. Other examples of ß-ketoesters useful for the production of chiral vicinal aminoalcohols include and ß-keto-3-phenylpropionic acid esters and ß-keto-4-phenylbutyric acid esters.

Central to this method for the production of these chiral vicinal aminoalcohols is the combination of two key steps, each of which proceeds with a well-defined and controllable stereochemical outcome. The first step is the stereoselective reduction of the keto group of a ß-ketoester to produce the corresponding ß-hydroxyester. This reaction is catalyzed by an alcohol dehydrogenase in the presence of a nicotinamide cofactor. Because of the rapid equilibrium between the enantiomers of a 2-substituted-ß-ketoester and the high degree of stereoselectivity of the dehydrogenase for reducing only one of the two enantiomers, two chiral centers are generated simultaneously. Therefore, this reaction provides for control of stereochemistry at both the C-2 and C-3 positions of the 2-substituted-ß-ketoester. The stereoselective reduction of the ß-ketoester may be carried out conveniently by biotransformation using any of a range of microorganisms, particularly yeast, that are able to reduce carbonyl groups in the presence of a carbon source such as glucose. By choosing a microorganism having the appropriate stereoselectivity for the reduction, this reaction can be used to generate a single diastereomer of the four possibilities in high optical purity. Alternatively, the stereoselective reduction may be carried out using an isolated alcohol dehydrogenase with recycling of the nicotinamide cofactor. This latter reaction has the advantage of well defined reaction conditions and the absence of multiple dehydrogenases that could lead to lower stereochemical purity of the product. Work is currently underway in our laboratory to identify dehydrogenases capable of producing each of the four possible diastereomers selectively.

In the example shown in Scheme 1. The ethyl ester of 2-ethylacetoacetate is reduced by the action of an alcohol dehydrogenase in the yeast *Saccharomyces cerevisiae* to produce ethyl (2R,3S)-2-ethyl-3-hydroxybutyrate. The reaction is highly diastereoselective, generating the (2R,3S) isomer in greater than 90% diastereomeric excess. (3)

After conversion of the chiral ß-hydroxyester to the corresponding hydroxamic acid by reaction with hydroxylamine, the second key step involves the rearrangement of the 2-substituted-3-hydroxy hydroxamic acid to the corresponding amino alcohol via the Lossen Rearrangement. This rearrangement occurs with retention of stereochemistry at the carbon bearing the carbonyl group, producing the desired in chiral vicinal aminoalcohol. Similarly, the chiral ß-hydroxyester can also be converted into the corresponding hydrazide by reaction with hydrazine, followed by Curtius Rearrangement to produce the chiral vicinal aminoalcohol. In the example in Scheme 1, Curtius Rearrangement leads to (3R,4S)-3-amino-4-hydroxypentane from the ethyl (2R,3S)-2-ethyl-3-hydroxybutyrate precursor.

Scheme 1: Production of (3R,4S)-3-amino-4-hydroxypentane

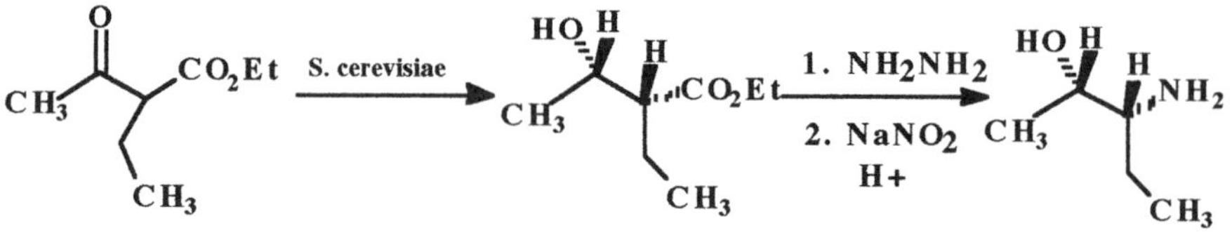

For production of the (3S,4S) stereoisomer, reduction of ethyl 2-ethylacetoacetate is carried out using the yeast *Geotrichum candidum*. The alcohol dehdyorgenase contained in this microorganism produces as its principal reduction product ethyl (2S,3S)-2-ethyl-3-hydroxybutyrate. (4) This 2-substituted-ß-hydroxyester is then converted to the corresponding hydrazide by reaction with hydrazine, followed by Curtius rearrangement to yield (3S,4S)-3-amino-4-hydroxypentane. The sequence of reactions is shown in Scheme 2.

Scheme 2: Production of (3S,4S)-3-amino-4-hydroxypentane

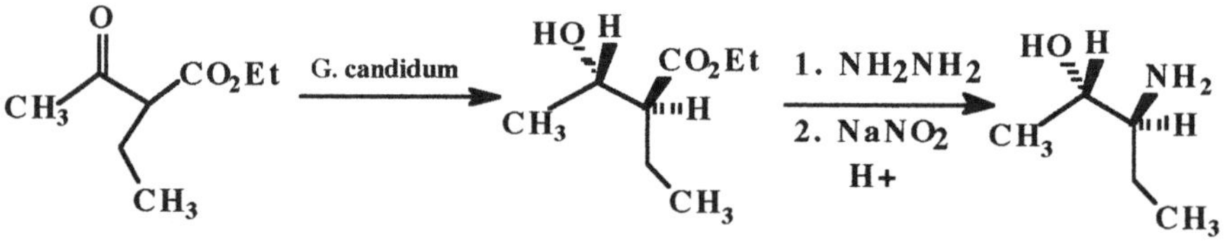

Scheme 3 shows an example using a cyclic ß-ketoester. In this case, the readily available ethyl 2-oxo-cyclohexanecarboxylate is used as a starting material. As before for acyclic precursors, by selecting the appropriate microorganism for the stereoselective reduction, chiral vicinal aminoalcohol products having the desired absolute configuration can be produced in high stereochemical purity. In this example, the yeast *Geotrichum candidum* produces only (1S,2S)-1 carboethoxy-2-hydroxycyclohexane by diastereoselective reduction.

Recycling of Nicotinamide Cofactors

As an alternative to using a whole cell biotransformation, the stereoselective reduction of the ß-ketoester can be carried out using an alcohol dehydrogenase. In this case, the nicotinamide cofactor must be recycled with an appropriate reductant to achieve

Scheme 3: Production of a Cyclic Chiral Aminoalcohol

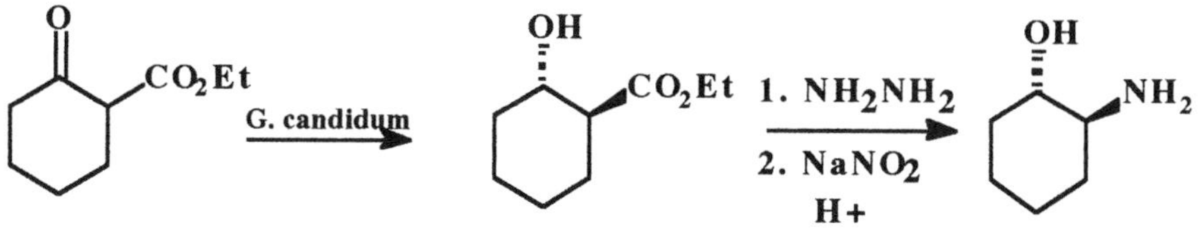

competitive economics. One of the most convenient recycle methods takes advantage of the reaction catalyzed by formate dehydrogenase, which converts NAD+ to NADH in the presence of formate. The overall reaction is illustrated in Scheme 4. Formate is an inexpensive reductant. Through optimization of reaction conditions to achieve high total recycle numbers for the nicotinamide cofactor, the use of an isolated alcohol dehydrogenase can be an attractive process option. By choosing a dehydrogenase having appropriate stereoselectivity, a desired 2-substituted-ß-hydroxyester can be produced in high chemical yields and high stereochemical purity. The 2-substituted-ß-hydroxyester is then converted to the chiral vicinal aminoalcohol by the same sequence of steps described previously.

The choice of alcohol dehydrogenase is critical to the success of the process described here. Dehydrogenases are widely distributed in nature and may be isolated from a large number of different microorganisms. One enzyme that has been described recently is the alcohol dehydrogenase from the microorganism *Candida parapsilosus*. (5) This enzyme catalyzes the stereoselective reduction of ß-ketoesters, including certain 2-substituted-ß-ketoesters. As an example, ethyl 2-methylacetoacetate is reduced in the presence of NAD+ to (2R,3S)-2-methyl-3-hydroxybutyrate with greater than 98% diastereomeric purity. (9) Similar results have also been obtained in our laboratory using the commercially-available alcohol dehydrogenase from *Rhodococcus erythropolis*. Currently, work is underway in our laboratory to expand the number of alcohol dehydrogenases that are available, and in particular, to identify additional alcohol dehydrogenases having different diastereoselectivity in the reduction of 2-substitited-ß-ketoesters.

Using either a whole cell biotransformation or an isolated alcohol dehydrogenase, the economics of this method are attractive. As an example, the chiral aminoalcohol 3-amino-4-hydroxypentane can be produced from the following simple and inexpensive chemical building blocks: ethyl 2-ethylacetoacetate (produced from ethyl acetoacetate and ethyl bromide), hydrazine, and inorganic acids and bases.

196

Scheme 4: Stereoselective reduction of a 2-Substituted-ß-ketoester Using an
Alcohol Dehydrogenase with Cofactor Recycling

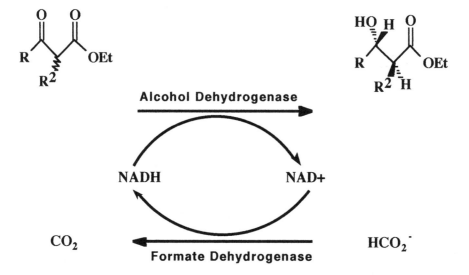

Experimental Examples of Stereoselective Reduction

Production of ethyl (2R,3S)-2-ethyl-3-hydroxybutyrate

Reduction is carried out according to the procedure of Seebach et al (7). In a representative reaction, twenty grams of bakers' yeast (Sigma Chemical Company, *Saccharomyces cerevisiae*, type II) is suspended in a solution of 30 grams of sucrose in water in a conical flask, and the mixture is placed in an orbital shaker chamber maintained at 220 rpm and 30 °C for 30 minutes to initiate fermentation. Two grams of ethyl 2-ethyl acetoacetate is dissolved in 2 ml of 95% ethanol, the resulting solution is added to the fermenting yeast, and shaking is resumed. The reaction is followed by TLC (staining with phosphomolybdic acid in ethanolic sulfuric acid) to monitor the consumption of starting material and the production of product alcohol. After approximately 48 hours the reaction is judged complete, and the reaction is terminated by removing the flask from the shaker and adding 20-30 grams of Celite to the reaction mixture. The resulting suspension is suction filtered through a pad of Celite, and the clear yellow filtrate is extracted with ethyl acetate (4 x 200 ml). The extracts are combined, dried over $MgSO_4$, filtered, and rotary evaporated to leave 1.6 grams of a yellowish oil containing ethyl (2R,3S)-2-ethyl-3-hydroxybutyrate as the major product (80%) and ethyl (2S,3S)-2-ethyl-3-hydroxybutyrate (20%) as the minor product.

Production of ethyl (2S,3S)-2-ethyl-3-hydroxybutyrate

Geotrichum candidum (ATCC 34614) is cultured according to the method of Buisson and Azerad (4) in one liter of a culture medium of glucose (30 grams), KH_2PO_4 (1 gram), K_2HPO_4 (2 grams), corn steep liquor (10 grams) $MgSO_4.7H_2O$ (0.5 gram), $NaNO_3$ (2 grams), $FeSO_4.7H_2O$ (0.02 gram), and KCl (0.5 gram) with rotary shaking at 25 °C. Two grams of ethyl 2-ethyl acetoacetate is dissolved in 2 ml of 95% ethanol, the resulting solution is added to the culture, and shaking is resumed. The reaction is followed by TLC (staining with anisaldehyde) to monitor the consumption of starting material and the production of product alcohol. After approximately 48 hours the reaction is judged complete, and the reaction is terminated by removing from the shaker and adding 20-30 grams of Celite. The resulting suspension is suction filtered through a pad of Celite and the clear yellow filtrate is extracted with ethyl acetate (4 x 200 ml). The extracts are combined, dried over $MgSO_4$, filtered, and rotary evaporated to leave 1.6 grams of (2S,3S)-2-ethyl-3-hydroxybutyrate as a yellow oil.

Production of ethyl (1S,2R)- 2-hydroxycyclohexanecarboxylate

Twenty-five grams of bakers' yeast (*Saccharomyces cerevisiae*, Sigma Chemical Company, type II) is suspended in 100 ml of sterilized tap water in a conical flask, and the mixture is placed on an orbital shaker (220 rpm) at 30 ^{o}C for 1 hour to activate the yeast. One gram of ethyl 2-oxocyclohexanecarboxylate is added, shaking is resumed, and the progress of the reaction is monitored by TLC (staining with anisaldehyde). After approximately 100 hours the reaction is judged complete, and the reaction is terminated by removing the flask from the shaker and adding 20-30 grams of Celite. The resulting suspension is suction filtered through a pad of Celite and the clear yellow filtrate is extracted with diethyl ether (4 x 100 ml). The extracts are combined, dried over $MgSO_4$, filtered, and rotary evaporated to leave 0.7 grams of ethyl (1R,2S)- 2-hydroxycyclohexanecarboxylate as a yellowish oil.

Production of ethyl (1S,2S)- 2-hydroxycyclohexanecarboxylate

Geotrichum candidum (ATCC 34614) is cultured according to the method of Buisson and Azerad (8) in one liter of a medium of glucose (30 grams), KH_2PO_4 (1 gram), K_2HPO_4 (2 grams), corn steep liquor (10 grams) $MgSO_4.7H_2O$ (0.5 gram), $NaNO_3$ (2 grams), $FeSO_4.7H_2O$ (0.02 gram), and KCl (0.5 gram) with rotary shaking at 25 ^{o}C. Two grams of ethyl 2-oxocyclohexanecarboxylate is dissolved in 2 ml of 95% ethanol, the resulting solution is added to the culture, and shaking is resumed. The reaction is followed by TLC (staining with anisaldehyde) to monitor the consumption of starting material and the production of product alcohol. After approximately 48 hours the reaction is judged complete, and the reaction is terminated by removing from the shaker and adding 20-30 grams of Celite. The resulting suspension is suction filtered through a pad of Celite and the clear yellow filtrate is extracted with ethyl acetate (4 x 200 ml). The extracts are combined, dried over $MgSO_4$, filtered, and rotary evaporated to leave 1.5 grams of ethyl (1S,2S)-2-hydroxycyclohexane-carboxylate as a yellow oil.

Conversion of ethyl (2S,3S)-2-ethyl-3-hydroxybutyrate to the hydrazide derivative

Ethyl (2S,3S)-2-ethyl-3-hydroxybutyrate (1 gram) is dissolved in 5 ml of absolute ethanol, followed by the addition of 0.5 gram of hydrazine. The solution is heated to reflux, and the progress of the reaction is followed by thin layer chromatography. After the reaction is judged complete, the ethanol is evaporated and the resulting residue redissolved in ethyl acetate. Hydrazine is removed by extraction with 1% HCl, and the ethyl acetate solution is dried over $MgSO_4$, filtered, and rotary evaporated to leave 0.9 grams of the hydrazide of (2S,3S)-2-ethyl-3-hydroxybutyrate.

Literature Cited

1. Rozzell, J. D. "Method for the Production of Chiral Vicinal Aminoalcohols," U. S. Patent 5,834,261 (1998).
2. Rozzell, J. D. "Precursors for the Production of Chiral Vicinal Aminoalcohols," U. S. Patent 5,942,644 (1999).
3. Physicians' Desk Reference, Edition 52; Medical Economics Company, Inc. 1998, pp. 1625-1628.
4. Physicians' Desk Reference, Edition 52; Medical Economics Company, Inc. 1998, pp. 2471-2475.
5. Physicians' Desk Reference, Edition 52; Medical Economics Company, Inc. 1998, pp. 459-464.
6. Physicians' Desk Reference, Edition 52; Medical Economics Company, Inc. 1998, pp. 476-480.
7. Seebach, D. M.; A. Sutter, A.; Weber, R. H. in *Preparative Biotransformations*, Roberts, S. M. Ed.; John Wiley & Sons Ltd. 1992; 2.1.1-2.1.5.
8. Buisson, D.; Azerad, R. *Tet. Lett.* **1986**, *27*, 2631-2634.
9. Kula, M.-R. Proceedings, IBC Congress on Enzyme Technology, March 1999.

Chapter 14

Stereoselectivity of Penicillin Acylase Catalyzed Ester Hydrolysis

Ravindra S. Topgi

R&D, Chemical Sciences, Pharmacia, 4901 Searle Parkway, Skokie, IL 60077

Penicillin G Amidohydrolase (penicillin acylase) is a unique enzyme that exhibits exceptionally high affinity to phenylacetic acid derivatives. This article deals with the structural limits of acceptability of phenylacetate esters as substrates and their influence on stereoselectivity of penicillin acylase catalyzed ester hydrolysis. The structural limits of alcohol moiety and their influence on stereoselectivity of penicillin acylase catalyzed ester hydrolysis are also considered.

Introduction

In comparison with non-biological based approaches, enzymatic syntheses have shown chemoselectivity, regioselectivity, enantioselectivity, rate enhancement and favorable environmental impacts.[1] Lipases[2] and esterases[3] catalyze the enantioselective hydrolysis of esters with good stereoselectivity. The remarkable selectivity (both regio and enantioselectivity) of amidases have been successfully utilized in several commercialized biocatalytic processes.[4] Penicillin G. Amidohydrolase (penicillin acylase) (E.C. 3.5.1.11) from *E. Coli* ATCC 9637 is a unique enzyme that exhibits exceptionally high affinity to phenylacetic acid derivatives.[5, 6] This capability has been used advantageously to achieve moderate to excellent stereochemical discrimination between corresponding enantiomers in the hydrolytic cleavage of the phenylacetyl group from α-aminoalkylphosphonic acids,[7, 8] α-, β- and γ-amino carboxylic acids,[9-11] sugars,[12] amines,[13] peptides[14] and esters

© 2001 American Chemical Society

of phenylacetic acid.[15, 16] Most of the acylases only accept α-amino acids. Penicillin acylase shows acceptability of wide range of substrates. The chemoselectivity of penicillin acylase for the phenylacetyl moiety has led to the development of new protecting groups for amino, hydroxy and carboxy functions, which have been used in peptide, carbohydrate and β-lactam chemistry.[17-19] Thus, penicillin acylase can act as an amidase as well as an esterase. Although this enzyme is extensively used in the hydrolysis of the amide bond, interesting exploration of its activity in the hydrolysis of phenylacetate esters encouraged us to document the recent advances in this report. This report is not intended to be a comprehensive review in the area, but discusses the recent advances in the hydrolysis of phenylacetic acid ester and its derivatives using penicillin acylase. In order to facilitate the discussion, the structure of ester has been divided into two parts - the phenylacetic acid moiety and the alcohol moiety.

Structural Features of Phenylacetic Acid Moiety

a. α-Substitution

The presence of α-substitution on the phenylacetyl side chain poses a question whether the enzyme penicillin acylase is able to recognize and stereodifferentiate the side chain according to its chirality. Fuganti et al. [15] studied the hydrolysis of a series of α-substituted phenylacetyl derivatives in order to explore the possibility of chiral discrimination using penicillin acylase.

Methyl mandelate **1a** was transformed at the highest rate, but the products obtained were of very low enantiomeric excess (ee). The methoxy derivative **2a** was hydrolyzed 50% in 2.5 hours, and both the acid and the recovered ester have an ee of 98%. Similarly, as in the preceding example, the (S)-enantiomer of formyloxy derivative **3a** was hydrolyzed first. Surprisingly, like compounds with larger substituents, i.e., **7a** and the acetate **8a** differing from formyloxy derivative **3a** by one methyl group, were not substrates. The structural requirement of an alkyl substituent in the alpha position is also strict. While the α-methyl derivative **4a** was hydrolyzed, the α-ethyl derivative **9a** was not hydrolyzed. Phenylacetyl esters bearing relatively bulky groups such as ethyl, acetyl or *t*-butyl are not accepted as substrates. The two halogen containing substrates, **5a** and **6a**, were not accepted efficiently by the enzyme. In these two cases, there is a difference in reactivity that can be correlated with the relative size and the electronic environment of the substituents. These

Table 1: Ester hydrolysis using penicillin acylase

a → Penicillin acylase → b + c

(% ee of Acids, (R) or (S))

Entry	X	R	R		R	
1	OH	Me	H	(15, (R))	Me	
2	OMe	Me	Me		H	(>98, (S))
3	OCHO	Me	Me		H	(>98, (S))
4	Me	Me	H	(64, (R))	Me	
5	Br	Me	H	(3, (R))	Me	
6	Cl	Me	H	(20, (R))	Me	
7	OTBDMSi	Me	Me		Me	
8	OAc	Me	Me		Me	
9	Et	Me	Me		Me	

SORUCE: Reproduced from Reference 15. Copyright 1992 Elsevier Science results indicate a very strict stereoelectronic requirement at the α-position. The striking difference due to polarity is evident in comparing the products obtained from **1a** and **2a**. The acids **2c** and **3c** have an absolute configuration opposite that in the ester **4c**. These results show that within the required phenylacetate structure, variation due to substituents in the α-position strongly influences the enantiomer recognition, which in turn influences the rate of hydrolysis and ee of the products.

b. Aromatic ring

Substituted phenyl:
Introduction of *p*-hydroxy, *p*-methoxy and *p*-amino groups on the phenyl ring have enhanced the rate of hydrolysis of amides by penicillin acylase. Cole used positional isomers of *p*-hydroxy phenylacetic acid in penicillin synthesis. In his study *p*-hydroxy isomer indicated high reactivity than the *m*-hydroxy isomer, which in turn is more reactive than *o*-hydroxy isomer.[20] Surprisingly, this type of study was never explored in the case of ester hydrolysis. However, the influence observed in penicillin acylase catalyzed amide hydrolysis may be applied to the ester hydrolysis.

Pyridine ring:
Waldmann et al. [21] in his pioneering work on substrate tuning, studied the influence of variations in the structure of the substrates in ester hydrolysis using penicillin acylase. This substrate structure variation introduced steric demand and/or polarity.[22] He referred this approach as "substrate tuning"- an approach that maps the substrate tolerance of a given enzyme.

In this '"substrate tuning" approach the phenyl group of phenylacetate was replaced with a pyridine ring. The presence of the nitrogen atom in the pyridyl acetic acid esters, recognized by penicillin acylase, should give rise to polar interactions that might lead to different binding within the active site of the enzyme. These polar interactions will be lacking when phenylacetates are employed. Pyridylacetic acid esters offer further substrate tuning with the possibility of placing the nitrogen atom in the *ortho-, para-,* or *meta-* position to the alkyl substituent and introducing charge on the nitrogen by alkylation or protonation. Solubility of pyridylacetic acid esters in aqueous solutions should enhance the velocity of the enzymatic transformation.

The relative velocities of enzymatic transformation (Table 2) indicate that the enzyme activity is strongly dependent on the position of the nitrogen atom in the aromatic ring. The *p*-pyridylacetate is hydrolyzed faster than *m*-pyridylacetate which is a better substrate than *o*-pyridylacetate.

Table 2: Hydrolysis of pyridylacetates using penicillin acylase

Entry	Substrate	Relative velocity*
10		97
11		76
12		44
13		8
14		44
15		42
16		39
17		4

* Hydrolysis of penicillin G 100% at pH 8

Table 3: Kinetic resolution of pyridylacetates using penicillin acylase

Entry	Substrate	Relative velocity*	Conversion %	ee** %
18		4	40	73 (R)
19		1.1	40	98 (R)
20		12	40	64 (S)
21		2.5	25	54 (R)
22		0.7	30	65 (R)
23		0.5	30	27 (R)

* Hydrolysis of penicilin G 100% at pH 8

** Enantiomeric excess and absolute configuration of the liberated alcohol

SORUCE: Reproduced from Reference 21. Copyright 1995 Elsevier Science

The racemic pyridylacetates **18-23** were subjected to enzymatic hydrolysis (Table 3) in order to study the influence of the nitrogen atom on the enantioselectivity of the enzyme. The pyridylacetate **18** gave (R)-1-phenylethanol with 73% ee at 40% conversion whereas under identical conditions, the corresponding phenylacetate gave only 28% ee. Similarly, pyridylacetate **19**, gave (R)-1-phenylpropanol with 98% ee at 40% conversion, whereas under identical conditions, the corresponding phenylacetate gave 90% ee at 35% conversion. (S)- and (R)-chiral alcohol enantiomers were obtained with 64% and 54% ee from pyridylacetates **20** and **21**, respectively. Under similar conditions, the corresponding phenylacetates gave the stereoisomers with 56% and 36% ee, respectively. Thus, the above results indicate that the introduction of a nitrogen atom into the phenyl ring of the phenylacetic acid moiety results in a significant increase in the enantioselectivity displayed by enzyme the penicillin acylase. In a trend similar to reaction velocity (Table 2), *p*-pyridylacetates are attacked with higher stereoselectivity than *m*-pyridylacetates, which are superior to *o*-pyridylacetates (Table 2: Compare entries **10**, **14**, and **17**). However, in comparison the *m*-pyridyl acetate gave a better result than the analogous phenyl acetate (65% ee vs. 28% ee), and only the *o*-pyridylacetate is comparable to the phenyl ester. Neither the hydrolysis at acidic pH, nor the use of N-methylpyridinium derivatives proved to be a valuable approach for enhancing either the velocity or stereodiscrimination of the enzymatic transformation.

Structural Features of Alcohol Moiety

a. Primary alcohol

Fuganti et al.[23] tried to establish the structural limits of acceptability of phenylacetate esters as substrates by penicillin acylase. Substrates structurally similar to penicillin G **24**, were chosen as the substrates.

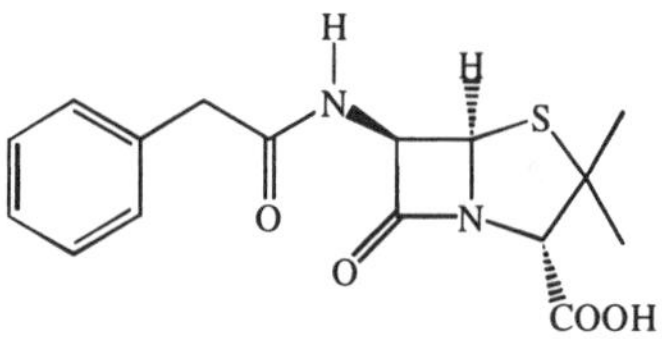

24

Figure 1: Penicillin G 24

The position of the geminal dimethyl group moiety of phenylacetates **25-33** is similar to Penicillin G **24**. The number of atoms that this moiety is separated from the bonds to be broken in the hydrolysis of the phenyl acetates **25-33** is the same as in the Penicillin G **24**.

Table 4: **Absolute configuration and enantiomeric excess values for the products obtained from corresponding phenylacetate ester at 50% hydrolysis using penicillin acylase**

Entry	R_1	R_2	R_3	R_4	R_5	Abs. configuration of the alcohol	ee %
25	H	H	H	Me	Me	(4S)	60
26	Me	H	H	Me	Me	(4S)	90
27	H	H	Me (4,5-anti)	Me	Me	(4S,5S)	50
28	H	H	CH$_2$OH (4,5-anti)	Me	Me	(4S,5S)	52
29	Me	Me	H (4,5-syn)	Me	Me	(4S,5R)	65
30	H	Me	Me	Me	Me	(4R)	33
31	H	H	Me (4,5-syn)	Me	Me	(4R,5S)	46
32	H	H	Et (4,5-syn)	Me	Me	(4R,5S)	80
33	Me	H	Et (4,5-syn)	Me	Me	(4R,5S)	90
34	H	H	H	Et	Et	(4S)	30
35	H	H	Et (4,5-syn)	Et	Et	(4R,5S)	40
36	H	H	H	Carbonyl		Optically inactive	-
37	H	H	Me (4,5-anti)	Carbonyl		Optically inactive	-

SORUCE: Reproduced from Reference 23. Copyright 1988 Elsevier Science

The absolute configuration of the products obtained in the hydrolysis and the enantiomeric excess (Table 4) strongly depend upon the nature of the 5-substituent(s) and the 4,5-stereochemistry. The highest ee (ca. 90%) values are obtained for the

4-methyl substituted dioxalane methanols derived from **26** and **33**. The absolute configuration at carbon 4 was inverted upon substitution of hydrogen at 5-position **34** with an ethyl group in *syn*-relationship **35**.

The influence of the 1,3-dioxolane moiety at the 2-position on enantioselective hydrolysis of phenyl esters was observed when hydrolysis of **34** and **35** was compared with their *gem*-dimethyl analogs **25** and **32**. At 50% hydrolysis **34** and **35** afforded (4S) and (4R, 5S) carbinols with 30% and 40% ee, respectively. Whereas **25** and **32** also afforded (4S) and (4R, 5S) carbinols but with 60% and 80% ee, respectively.

Substitution of a sp^2 carbonyl carbon for a sp^3 carbon bearing two methyl groups in position 2 of the dioxolane structure, as was the case in cyclic carbonates **36** and **37**, resulted in rapid hydrolysis (t ½ ca. 1 h). The products obtained at 50% hydrolysis were devoid of optical activity.

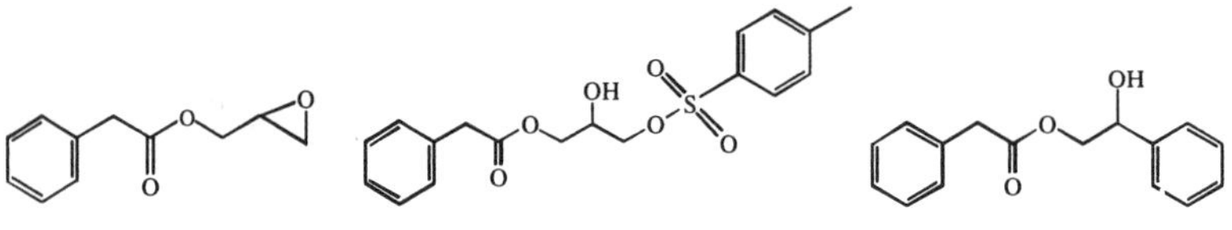

Figure 2: Phenylacetate esters of primary alcohols

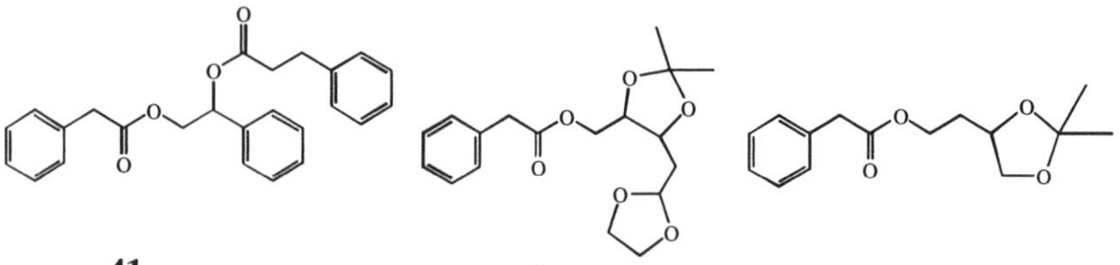

The glycidol (2,3-epoxypropan-1-ol) phenylacetate **38**, an analog of **25**, is rapidly hydrolyzed. At 50% hydrolysis the recovered ester was optically inactive. A similar result was obtained in penicillin acylase hydrolysis for another glycerol derivative **39**. In the case of the 2-phenylacetate of 1-phenyl-1,2-ethanediol **40**, the 50% hydrolysis afforded the alcohol with poor optical purity. When the alcohol group of the 2-phenylacetate of 1-phenyl-1,2-ethanediol **40** was esterified with phenylpropionic acid, the diester **41** was not hydrolyzed by penicillin acylase. Phenylacetate ester **42** was hydrolyzed to 50% to give a carbinol (4R,5S) with 70% ee. Similarly, the hydrolysis of the phenylacetate ester of 3,4-isopropylidenedioxybutan-1-ol **43** gave a carbinol (S) with 28% ee. The comparison of these results with those for the lower analog **25** indicated a decreased optical purity on introducing an additional CH_2 group between the asymmetric carbon atom and the bond to be broken.

When α-chymotrypsin was used to hydrolyze the phenylpropionate esters of the same carbinols, the products in general showed poor ee values and an absolute stereochemistry opposite to that observed in the penicillin acylase hydrolysis of phenylacetyl esters. [24]

Waldmann et al.[21] used pyridylacetates of primary alcohols to study the substrate tuning in penicillin acylase hydrolysis. In the case of pyridylacetates **10, 11, 14, 15,** and **17**, the relative reaction velocity of the enzymatic hydrolysis declined with increasing steric hindrance of the alcohol (Table 2). The relative velocities of pyridylacetates reveal that the enzyme activity is strongly dependent on the position of the nitrogen atom in the aromatic ring.

b. Secondary Alcohols

Fuganti et al.[23] studied the ester hydrolysis of secondary alcohols using penicillin acylase. In his studies the framework of the dioxolane methanols **25, 27,** and **31** was substituted in the α-position with a vinyl group, thus mimicking in some way in terms of number of atoms and electronic density, the carbonyl group of the β-lactam ring of penicillin G **24**.

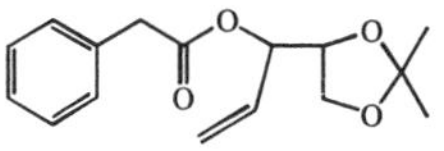

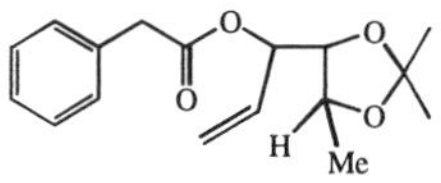

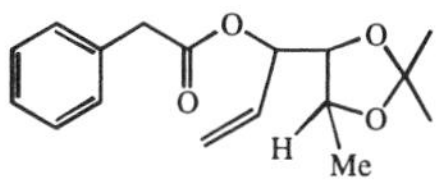

Figure 3: Phenylacetate esters of secondary alcohols

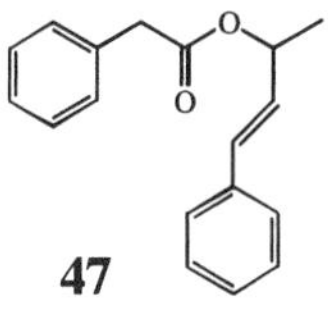

However, the resulting respective secondary allylic esters **44-46** were not hydrolyzed. On the other hand, penicillin acylase does catalyze the hydrolysis of the β-phenyl substituted secondary allylic phenylacetate **47**, yielding with a (2R) carbinol of 80-85% ee.

Fuganti et al. [15] studied the presence an α-substitution on the phenylacetyl esters of secondary alcohols to explore the possibility of chiral discrimination using penicillin acylase.

Table 5: Hydrolysis of phenylacetate esters of secondary alcohols using penicillin acylase

Entry	X	R	% ee of Acid ((R) or (S))		R
1	OH	Me	H	(15, (R))	Me
48	OH	i-Pr	H	(20, (R))	i-Pr
4	Me	Me	H	(64, (R))	Me
49	Me	i-Pr	H	(>98, (R))	i-Pr

SORUCE: Reproduced from Reference 15. Copyright 1992 Elsevier Science

Methyl mandelate **1a** was transformed at the highest rate, but the products obtained were of very low ee. The isopropyl ester **48** was recovered with about 20% ee. The hydrolysis of methyl ester **4a** gave the methyl ester **4c** with an enantiomeric excess 76% at 25% conversion, but lower enantiomeric excess 64% at 50% conversion. However, the isopropyl ester **49** was hydrolyzed giving both the acid and the unreacted ester in >98% ee. Thus, passing from methyl ester to isopropyl ester there is a significant increase in the ee value.

Baldaro et al. [16] studied the phenylacetates of secondary alcohols with the aim of obtaining the products enriched in one enantiomer. The compounds bearing an α-carboxylate were found to be particularly good substrates for the enzyme penicillin acylase. This study indicates the selective hydrolysis of phenylacetate group in presence of methyl or ethyl esters.

For example, phenylacetates of α-hydroxy esters **50**, **51** and **52** are efficiently hydrolyzed to give products with high ee. Small structural modifications dramatically change the reaction rate and/or the enantiospecificity of the reaction **52-54**. In comparison with other substrates the very low enantiomeric excess of compounds **55c** and **56c** can be attributed to the absence of the aromatic ring as a part of the alcohol moiety. Thus, an aromatic moiety seems beneficial for a good fit into the enzyme binding site.

Table 6: Penicillin acylase catalyzed hydrolysis of phenylacetates of secondary alcohols

	% Hydrolysis	Alcohol	ee, %
	50		26
	50		10
	50		40
	50		56
	12		94
	50		92
	25		94
	50		90
	18		>98
	32		36

SORUCE: Reproduced from Reference 16. Copyright 1993 Elsevier Science

Table 7: Penicillin acylase catalyzed hydrolysis of pyridylacetates of secondary alcohols

Entry	Substrate	Relative velocity*	Conversion %	ee** %
60		44	-	-
61		8	-	-
62		39	-	-
63		4	40	73 (R)
64		1.1	40	98 (R)
65		0.7	30	65 (R)
66		0.5	30	27 (R)

* Hydrolysis of penicilin G 100% at pH 8

** Enantiomeric excess and absolute configuration of the liberated alcohol

SORUCE: Reproduced from Reference 21. Copyright 1995 Elsevier Science

Waldmann et al.[21] used pyridylacetates of secondary alcohols to study the substrate tuning in penicillin acylase hydrolysis.

An extension of the carbon chain by one carbon dramatically lowers the relative reaction velocity from 44 (**60**) to 8 (**61**). Thus, the relative velocity of the enzymatic hydrolysis of pyridylacetates **60-66** declines with increasing steric hindrance of the secondary alcohol moiety. Comparison of the enzymatic hydrolysis of pyridylacetates **63-66** reveals the influence of the position of the nitrogen atom in the aromatic ring on the relative reaction velocity. Thus, similar to the primary alcohols, the relative velocities of pyridylacetates reveal that the enzyme activity is strongly dependent on the position of the nitrogen atom in the aromatic ring.

Waldmann et al.[25] synthesized enantiomerically enriched 2-furylcarbinol derivatives with useful ee's by kinetic resolution of the respective phenylacetates **67-69** using penicillin acylase. He applied this approach in the synthesis of enantiomerically enriched L-aculose and L-cinerulose starting from 2-furyl methylcarbinol.

Table 8: Penicillin acylase catalyzed hydrolysis of phenylacetates of 2-furylcarbinols

Entry	R	Relative velocity* %	Enantiomer ratio of recovered material	Configuration of fastrer attacked enantiomer
67	COOMe	3.4	91 : 9	(S)
68	CN	6.2	86 : 14	(R)
69	CH_3	2.7	90 : 10	(R)

* Hydrolysis of benzylpenicillin 100% at pH 7.5

SORUCE: Reproduced from Reference 25. Copyright 1989 Elsevier Science

The enzyme tolerated the presence of different functional groups adjacent to the ester moiety and afforded the desired furan derivatives with useful ee's.

The stereoelectronic influence of the substituent alpha to the furan ring can be seen in the relative velocity of the enzymatic hydrolysis. The substitution of an amino group with the hydroxy group increases dramatically the enantiomer ratio.[26]

The utility of the penicillin acylase for the resolution of secondary alcohols seems to be ruled by more subtle requirements in terms of steric and polar character than with most of the widely used hydrolytic enzymes.

Conclusion

The ability of the enzyme penicillin acylase to exhibit esterase and amidase activity under different reaction conditions offers development of new techniques and its application to biotransformation and manipulation of protecting groups. This application of penicillin acylase can advantageously complement the stereoselectivity offered by the presently used lipases and esterases. The efficiency of penicillin acylase catalyzed ester hydrolysis reactions can be enhanced significantly in terms of reaction velocity and stereoselectivity by choosing the appropriate phenylacetyl derivative.

Because of the widespread availability of lipases, esterases and Baker's yeast to organic chemists, chemoenzymatic synthetic approaches using these enzymes will continue. The use of penicillin acylases will continue to be of major interest because of their wide substrate tolerability and the maturity of these enzymes in industrial use worldwide compared to asymmetric syntheses for scaleup. The advance of genomics, the emergence of thermostable enzymes and advanced enzyme engineering technology such as directed evolution and DNA shuffling will help to propel the use of enzymes in large scale biocatalysis processes. Their potential for new applications and the rationalization of their properties is a good challenge for chemists and biochemists.

Acknowledgments

I gratefully acknowledge James R. Behling, John Ng, John Dygos and Kathy McLaughlin for valuable discussions and manuscript preparation.

References

1. Jones, J.B. *Tetrahedron*, **1986**, *42*, 3351.
2. *Enzyme Catalysis in Organic Synthesis*; Drauz, K.; Waldmann, H., Eds; VCH: New York, **1995**; pp 178-195
3. Patel, R. N.; Banerjee, A.; Howell, J. M.; McNamee, C. G.; Brozozowski, D.; Mirfakhrae, D.; Nanduri, V.; Thottathil, J. K.; Szarka, L. *Tet. Asymm.* **1993**, *4*, 2069.
4. *Enzyme Catalysis in Organic Synthesis*; Drauz, K.; Waldmann, H., Eds; VCH: New York, **1995**; pp 386.
5. Didziapetris, R.; Drabnig, B.; Sehellenberger, V.; Jakubke, H.-D.; Svedas, V. *FEBS Lett.* **1991**, *287*, 31.
6. Stoineva, I. B.; Galunsky, B. P.; Lazanov, V. S.; Ivanov, I. P.; Petkov, D. D. *Tetrahedron*, **1992**, *48*, 1115.
7. Solodenko, V. A.; Kasheva, T. N.; Kukhar, V. P.; Kozlova, E. V.; Mironenko, D. A.; Svedas, V. K. *Tetrahedron*, **1991**, *47*, 3989.
8. Solodenko, V. A.; Belik, M. Y.; Hgalushko, S. V.; Kukhar, V. P.; Kozlova, E. V.; Mironenko, D. A.; Svedas, V. K. *Tet. Asymm.* **1993**, *4*, 1965.
9. Margolin, A. L.; Svedas, V. K.; Berezin, I. V. *Biochim. Biophys. Acta*, **1980**, *616*, 283.

10. Anderson, E.; Mattiasson, B.; Hahn-Hagerdal, B. *Enz. Microb. Technol.* **1984**, *6*, 301.
11. Margolin, A. L. *Tet. Lett.* **1993**, *34*, 1239.
12. Waldmann, H. *Kontakte*, **1991**, *2*, 33.
13. Romeo, A.; Lucente, G.; Rossi, D.; Zanotti, G. *Tet. Lett.* **1971**, *21*, 1799.
14. Svedas, V. K.; Galaev, I. Yu.; Semiletov, Yu. A.; Korshunova, G. A. *Bioorgan. Khim.* **1983**, *9*, 1139.
15. Fuganti, C.; Rosell, C. M.; Servi, S.; Tagliani, A.; Terreni, M. *Tet. Asymm.* **1992**, *3*, 383.
16. Baldaro, E.; D'Arrigo, P.; Pedrocchi-Fantoni, G.; Rosell. C. M.; Servi, S.; Tagliani, A.; Terreni, M. *Tet. Asymm.* **1993**, *4*, 1031.
17. Fuganti, C.; Grasselli, P.; Casati, P. *Tet. Lett.*, **1986**, *27*, 3191.
18. Baldaro, E.; Faiardi, D.; Fuganti, C.; Grasselli, P.; Lazzarini, A. *Tet. Lett.*, **1988**, *29*, 4623.
19. Hermann, P. *Biomed. Biochim. Acta*, **1991**, *50*, 19.
20. Cole, M. *Biochem. J.*, **1969**, *115*, 747.
21. Pohl, T.; Waldmann, H. *Tet. Lett.* **1995**, *36*, 2963.
22. *Biotransformations in organic chemistry*, Faber, K. Springer-Verlag, Berlin Heidelberg **1992**.
23. Fuganti, C.; Grasselli, P.; Servi, S.; Lazzarini, A.; Casati, P. *Tetrahedron*, **1988**, *44*, 2575.
24. Fuganti, C.; Grasselli, P.; Servi, S.; Lazzarini, A.; Casati, C. *J. Chem. Soc., Chem. Commun.* **1987**, 538.
25. Waldmann, H. *Tet. Lett.* **1989**, *36*, 3057.
26. Baldaro, E.; Fuganti, C.; Servi, S.; Tagliani, A.; Terreni, M. In *Microbial Reagents in Organic Synthesis* Servi, S. Ed,; Kluwer Academic Publishers: Netherlands, **1992**, 175-188.

Chapter 15

Synthesis of Chiral Pharmaceutical Intermediates by Oxidoreductases

Ramesh N. Patel and Ronald L. Hanson

Department of Enzyme Technology, Process Research, Bristol-Myers Squibb Pharmaceutical Research Institute, P.O. Box 191, New Brunswiick, NJ 08903

Chiral intermediates were prepared by enzymatic process using oxidoreductases for the chemical synthesis of pharmaceutical drug candidates. These includes: (1) the microbial reduction of 1-(4-fluorophenyl)-4-[4-(5-fluoro-2-pyrimidinyl)-1-piperazinyl]-1-butanone **1** to R-(+)-1-(4-fluoro-phenyl)-4-[4-(5-fluoro-2-pyrimidinyl)-1-piperazinyl]-1-butanol [R-(+)-BMY 14802], a antipsychotic agent; (2) the reduction of N-(4-(1-oxo-2-chloroacetyl ethyl) phenyl methane sulfonamide **3** to corresponding chiral alcohol **4**, an intermediate for D-(+)-N-[4-[1-Hydroxy-2-[(-methylethyl)amino]ethyl]phenyl] methanesulfonamide [D-(+) sotalol], a β-blocker with class III antiarrhythmic properties; (3) biotransformation of Nε-carbobenzoxy (CBZ)-L-lysine **7** to CBZ-L-oxylysine **5** an intermediate needed for synthesis of (S)-1-[6-amino-2-[[hydroxy(4-phenylbutyl) phosphinyl]oxy]1-oxohexyl]-L-proline [ceronapril] , a new angiotensin converting enzyme [ACE] inhibitor **6** (4) enzymatic synthesis L-β-hydroxyvaline **9** from α-keto-β-hydroxy isovalerate **16**. L-β-Hydroxy valine **9** is a key chiral intermediate needed for the synthesis of [S-(Z)]-[[[1-(2-Amino-4-thiazolyl)-2- [[2,2-dimethyl-4-oxo-1-(sulfooxy)-3-azetidinyl] amino]-2-oxoethylidene] amino]oxy]acetic acid [tigemonam] **10** , a orally active monobactam, (5) enzymatic synthesis of L-6-hydroxynorleucine **17**, and (6) enzymatic synthesis of (S)-2-amino-5-(1,3-dioxolan-2-yl)-pentanoic acid (allysine ethylene acetal, **21**), one of three building blocks used for an alternative synthesis of omapatrilat, a vasopeptidase inhibitor.

Introduction

Recently much attention has been focused on the interaction of small molecules with biological macromolecules. The search for selective enzyme inhibitors and receptor agonists or antagonists is one of the keys for target-oriented research in the pharmaceutical industry. Increasing understanding of the mechanism of drug interaction on a molecular level has led to the wide awareness of the importance of chirality as the key to the efficacy of many drug products. It is now known that in many cases only one stereoisomer of a drug substance is required for efficacy and the other stereoisomer is either inactive or exhibits considerably reduced acitivity.

Pharmaceutical companies are aware that, where appropriate, new drugs for the clinic should be homochiral to avoid the possibility of unnecessary side effects due to an undesirable stereoisomer. In many cases where the switch from racemate drug substance to enantiomerically pure compound is feasible, there is the opportunity to extend the use of an industrial process. The physical characteristic of enantiomers versus racemates may confer processing or formulation advantages.

Chiral drug intermediates can be prepared by different routes. One is to obtain them from naturally occuring chiral synthons mainly produced by fermentation processes. The chiral pool primarily refers to inexpensive, readily available, optically active natural products. Second is to carry out the resolution of racemic compounds. This can be achieved by preferential crystallization of stereoisomers or diastereoisomers and by kinetic resolution of racemic compounds by chemical or biocatalytic methods. Finally, chiral synthons can also be prepared by asymmetric synthesis by either chemical or biocatalytic processes using microbial cells or enzymes derived therefrom. The advantages of microbial or enzyme catalyzed reactions over chemical reactions are that they are stereoselective, can be carried out at ambient temperature and atmospheric pressure. This minimizes problems of isomerization, racemization, epimerization and rearrangement that generally occur during chemical processes. Biocatalytic processes are generally carried out in aqueous solution. This will avoid the use of environmentally harmful chemicals used in the chemical processes and solvent waste disposal. Furthermore, microbial cells or enzymes derived therefrom can be immobilized and reused many cycles.

Recently, a number of review articles (1-9) have been published on the use of enzymes in organic synthesis. This report provides some specific examples of the use of oxidoreductases in stereoselective catalysis and preparation of chiral drug intermediates required for our antipsychotic, antiarrhythmic (β-Blocker), antihypertensive and antibacterial agents (10-13).

Antipsychotic Drug(BMY-14802)
Stereoselective Microbial Reduction of 1-(4-Fluorophenyl)-4-[4-(5-Fluoro-2-Pyrimidinyl)-1-Piperazinyl]-1-Butanone 1.

During the past few years, much research effort has been directed towards the understanding of the Sigma receptor system in brain and endocrine tissue. This effort has been motivated by the hope that the Sigma site may be a target for a new class of antipsychotic drugs (14-17). The characterization of the Sigma system helped to clarify the biochemical properties of the distinct haloperidol-sensitive Sigma binding site, the pharmacological effects of Sigma drugs in several assay systems, and the transmitter properties of a putative endogenous ligand for the Sigma site (18-21).

R (+) compound 2 (BMY 14802) is a Sigma ligand and has a high affinity for Sigma binding sites and can selectively inhibit conditioned avoidance and apomorphine-induced stereotype in rats predictive of antipsychotic efficacy (22,23). The dopamine cell firing in the *Substantia Nigra* caused by the putative Sigma receptors was inhibited by (+)-3H-3-(3-hydroxyphenyl)-N-(1-propyl) piperidine (17, 18). In this section, we are describing the stereoselective microbial reduction of 1-(4-fluorophenyl)-4-[4-(5-fluoro-2-pyrimidinyl)-1-piperazinyl]-1-butanone 1 to yield R (+) BMY 14802 2 [Scheme 1]

218

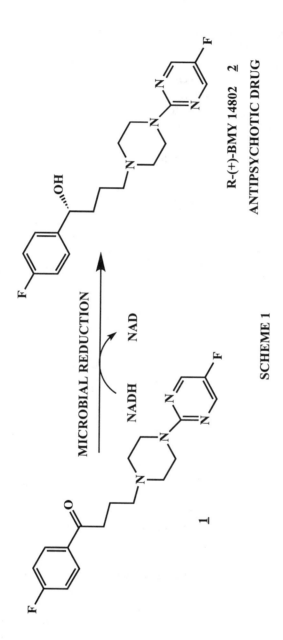

MICROBIAL REDUCTION

NADH   NAD

1

R-(+)-BMY 14802   2
ANTIPSYCHOTIC DRUG

SCHEME 1

Among microorganisms evaluated for the reduction of compound **1** to **2**, *M. ramanniana* ATCC 38191 predominately reduced compound **1** to R (+) compound **2** and *Pullularia pullulans* ATCC 16623 reduced compound **1** to S (-) compound **2**. The enantiomeric excess (ee) of > 98% was obtained in each reaction (Table 1).

Further research was conducted using *M. ramanniana* ATCC 38191 to convert compound **1** to R (+) compound **2**. Cells of *M. ramanniana* ATCC 38191 were grown in a 280-L fermentor containing 250-L of medium. During growth, cells were harvested periodically, and used to conduct the reduction of compound **1** by cell suspensions of microorganisms to determine optimum time to harvest cultures. As shown in the Table 2, cells harvested after 31 hours growth and used in the bioreduction of compound **1** gave higher reaction yield (89%) and the ee (> 98%) of R (+) compound **2**.

Cells harvested from a 380-L fermentor were evaluated for the reduction of compound **1** in a 5-L and 15-L fermentors using 3-L and 10-L cell-suspensions (20% w/v, wet cells), respectively. Compound **1** was supplied at 2 g/L concentration. Glucose was supplemented at 70 g/L concentration. After a 24-hour reaction period, about 99% yield (99 % ee) of R (+) compound **2** was obtained from both batches. The kinetics of transformation of compound **1** are as shown in Table 3. Isolation of R (+) compound **2** from the 10-L fermentation broth gave 14.2 gram of product in overall 70 M% yield and 99% HPLC area % purity. Isolated R (+)-compound **2** gave a melting point of 115°C, specific rotation $\alpha[D]_{25}$ of +26.8° using chloroform as solvent and ee of 99 % as analyzed by chiral HPLC.

The effect of pH and temperature on the reduction of compound **1** to R (+)-compound **2** by *M. ramanniana* ATCC 38191 was evaluated in a 100 ml reactor. The optimum pH for the reduction of compound **1** to R (+) compound **2** is 5.5. The optimum temperature for the reduction of compound **1** to R (+) compound **2** is 28°C. The effect of glucose and substrate concentration on the biotransformation of compound **1** to R (+) compound **2** was evaluated. 20 mg/ml of glucose was enough to get a 98% reaction yield. In the absence of added glucose, only 22% reaction yield was obtained (Table 4).

A single-stage fermentation/biotransformation process was developed for conversion of compound **1** to compound **2** by cells of *M. ramanniana* ATCC 38191. Cells were grown in a 20-L fermentor containing 15 L of medium A and after 40 hours of growth in a fermentor, when the residual glucose was depleted (0.1%) the pH of the medium dropped to 4.5 compound **1** (2 g/L) and glucose (70 g/L) were added to the fermentor and the biotransformation process was continued (Figure 1). The biotransformations process was completed in a 24-hour period, with the reaction yield of 99% and the ee of 98 % for R (+) compound **2**. At the end of biotransformation process, cells were removed by filtration and product was recovered from the filtrate. Filtrate was adjusted to pH 8.0 and allowed to stand at room temperature for 1 hour. The precipitate was filtered and dissolved in isopropyl alcohol (IPA). Product was precipitated as a HCl salt by addition of isopropanolic HCl. Recrystallization from IPA yielded 24 grams of R (+) compound **2**. Overall, 80% recovery of product was obtained in 99% HPLC purity and 98% ee.

Table I. Microbial Reduction of 1 to 2 [BMY 14802]

Microorganisms	Substrate 1 (g/L)	BMY 14802 (g/L)	EE of BMY 14802	
			R-(+)	S-(-)
Mortierella ramanniana ATCC 38191	0.02	1.95	98.9	1.1
Pullularia pullulans ATCC 16623	0.04	1.62	1.5	98.5

Microorganisms were grown in a 25-L fermentor for 48 hours. Cells were harvested by Cepa centrifuge and suspended in 0.1 M phosphate buffer pH 6.0. Cell suspensions (10% w/v, wet cells) were used in the reduction of 1 (2 g/L) at 280 RPM and 28°C for 24 hours. The concentrations of BMY-14802: 1-(4-fluorophenyl)-4-[4-(5-fluoro-2-pyrimidinyl)-1-piperazinyl]1-butanol.

**Table II. Reduction of 1 to [BMY 14802] by M. ramanniana:
Evaluation of cells during growth in a 380 L Fermentor**

Cells Harvest time (hours)	Substrate 1 Concentration (g/L)	BMY 14802 Concentration (g/L)	EE of R-(+) BMY 14802 (%)
7	0.96	0.42	89.5
19	0.56	1.1	93.3
31	0.04	1.86	99.4
43	0.12	1.8	96.8

During fermentation cells were harvested from 500 ml broth samples and suspended in 100 mM potassium phosphate buffer (pH 5.8). Cell suspensions (20% w/v, wet cells) were used to conduct the reduction of 1 (2 mg/ml) at 280 RPM and 20 C for 24 hours. Concentration of 1 and 2 were determined by GC and the optical purity of 2 was determined by chiral HPLC. BMY-14802: see table 1 legend.

Table III. Reduction of 1 to 2 by cell suspensions of M. ramanniana
ATCC 38191: Preparative 10 liter batch.

Reaction time (hours)	Substrate 1 Concentration (g/L)	BMY 14802 Concentration (g/L)	Conversion (%)	EE of R-(+) BMY 14802 (%)
3	1.65	0.31	15	ND
6	1.3	0.64	32	ND
12	0.71	1.25	62	ND
18	0.48	1.61	80	99.4
24	0.04	1.98	99	99.5

Cells of M. ramanniana ATCC 38191 were suspended in 10 liter of 0. 1 M potassium phosphate buffer (pH 8) at 20% (w/v, wet cells) concentration. Glucose (200 gram) and 1 (20 gram) were supplied and the reduction was conducted in a 15-liter fermentor at 500 RPM, 28 C, 12 LPM aeration. BMY-14802: see table 1 legend.

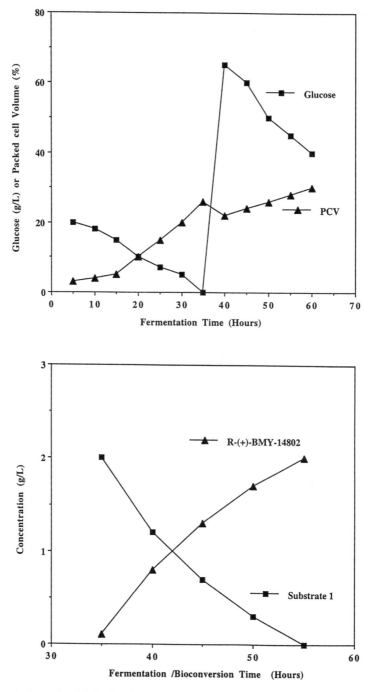

Figure 1: Growth of Mortierella rammaniana in a 15 L fermenot during single-stage fermentation/biotransformation process for reduction of 1 to R-(+)-BMY-14802.

The enzyme which reduces **1** to R (+) compound **2** was purified 55 fold to homogeneity with a specific activity (nmoles of R (+) compound **2** formed per minute per mg of protein) of 268. Protein concentration was determined by Bradford's method (24). The purified enzyme has a molecular weight of 29,000 daltons as determined by gel filtration and consists of a single subunit of 29,000 daltons as judge from sodium dodecyl sulfate polyacrylamide gel electrophoresis. The purified enzyme was inhibited by p-hydroxy mercuribenzoate, 1,10-phenanthroline, and 2,2-bipyridyl at 1 mM concentration of each inhibitor. This indicates the involvement of a sulfydryl group (and perhaps metals) at the enzyme active site.

Antiarrhythmic Agent (β-Blocker)
Reduction of N-(4-(1-Oxo-2-Chloroacetyl Ethyl) Phenyl Methane Sulfonamide **3** to Corresponding Chiral Alcohol **4**.

Larsen and Lish (25) reported the biological activity of a series of phenethanolamine bearing alkyl sulfonamido groups on the benzene ring. Within this series, some compounds possesed adrenergic and antiadrenergic actions including α-adrenergic receptor block or receptor stimulation, β-adrenergic receptor block or receptor stimulation. D-(+) sotalol [Scheme 2] is a β-blocker (26). Unlike other β-blocker it has class III antiarrhythmic properties (27) . The β-adrenergic blocking drugs such as propranolol [(1-[(-Methylethyl) amino]-3-(1-naphthalenyloxy)-2-propanol] and sotalol have been separated chemically into the dextro and levo rotatory optical isomers, and it has been demonstrated that the activity of the levo isomer is 50 times that of the corresponding dextro isomer (28). In this section, We are describing the stereoselective microbial reduction of N-(4-(2-chloroacetyl)phenyl)methane-sulfonamide **3** to the corresponding (+)- **4**. The compound (+)-**4** is a key chiral intermediate for the synthesis of D-(+)- sotalol.

Microorganisms were screened for the transformation of compound **3** to compound **4**. As shown in Table 5, among cultures evaluated *Rhodococcus sp.* ATCC 29675, ATCC 21243, *Nocardia salmonicolor* SC 6310 and *Hansenula polymorpha* ATC C 26012 gave desired product **4** in >90% optical purity. *Hansenula polymorpha* ATCC 26012 catalyze the efficient conversion of compound **3** to compound **4** in 95% reaction yield and >98% ee.

Growth of *H. polymorpha* ATCC 26012 culture was carried out in a 380 L fermentor. Cells harvested from fermentor were used to conduct transformation in a 3 L preparative batch. Cells were suspended in 3 L of 10 mM potassium phosphate buffer pH 7.0 to 20% (w/v, wet cells) cell concentration. Compound **3** (12 grams) and glucose (225 grams) were added to the fermentor and the reduction reaction was carried out at 25°C, 200 RPM, pH 7. About 95% conversion of compound **3** to compound **4** was obtained in 20 hours reaction period (Table 6). The isolation of compound **4** from the reaction mixture was carried out using a preparative HPLC to obtain 8.2 grams of product. The isolated **4** gave a specific rotation $[\alpha]_D$ Na of +20° and ee of >98% as analyzed by chiral HPLC. The Mass spectra and H-NMR spectra of isolated compound **4** and standard compound **4** were virtually identical. Reduction of **3** to **4** was also carried out using cell extract of *H. polymorpha* ATCC 26012. Formate dehydrogenase was used to regenerate cofactor NADH required for reduction

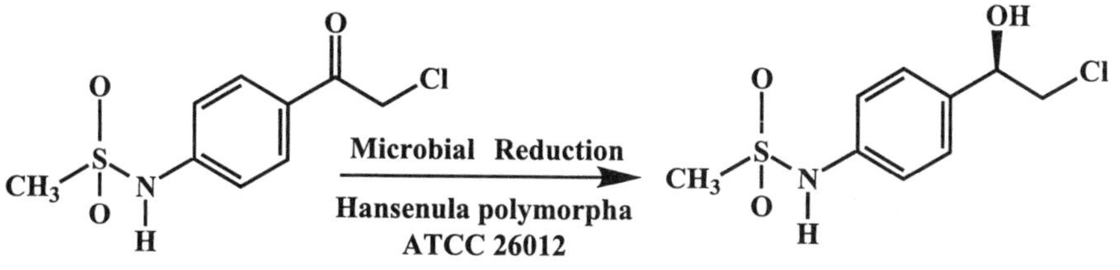

Compound 3 (+)-Compound 4

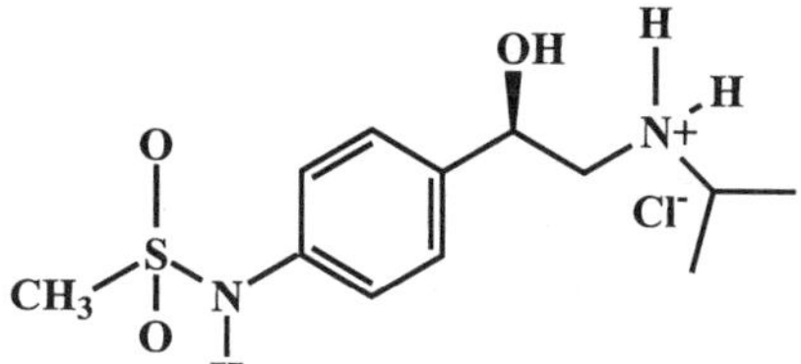

D-(+) Sotalol HCl

ANTIARRHYTHMIC AGENT (β-BLOCKER)

SCHEME 2

Table IV. Reduction of 1 to BMY-14802 by cell suspensions of M. ramanianna ATCC 38191: Effect of glucose concentration

Glucose added (g/L)	BMY 14802 (g/L)	Conversion (%)	EE of R-(+) BMY 14802 (%)
0	0.48	24	
10	1.49	75	99.4
20	1.95	98	99.4
40	1.93	97	99.3

Cells suspensions of M. ramanniana ATCC 38191 (8% w/v, wet cells) were supplied with 1 (2 mg/ml) and glucose as indicated. Reactions were conducted at pH 5.5, 280 RPM, 28 C, for 20 hours. Compound 1 and 2 were analyzed by GC and the optical purity of 2 was determined by chiral HPLC. BMY-14802: see table 1 legend.

Table V. Microbial Reduction of N-(4-(1-oxo-2-chloroacetyl ethyl)
phenyl methane sulfonamide 3

Microorganisms	Reaction time (hours)	Conversion to alcohol 4 (%)	EE of alcohol 4 (%)
Pichia methanolica ATCC 58403	40	80	26
Pullularia pullulans ATCC 16623	68	73	75
Hansenula polymorpha ATCC 26012	22	95	99
Trichoderma polysporium SC 14962	22	86	77
Rhodococcus sp. ATCC 29675	22	53	91
Rhodococcus sp. ATCC 21243	22	64	97.5
Nocardia salmonicolor SC 6310	22	45	99

Cell suspensions (20% w/v of wet cells) of each microbial cultures in 10 ml of 10 mM phosphate buffer were supplemented with 20 mg of substrate and 750 mg of glucose. Bioreduction of 3 was carried out at 28° C, and 200 RPM on a rotary shaker. The concentration of substrate 3 and product 4 were determined by GC and the optical purity of 4 was determined by chiral HPLC using a chiralcel OB column.

reaction. After 18 hours reaction time, 95% conversion of **3** to **4** was obtained at 2 g/L substrate concentration.

Antihypertensive Drug

Biotransformation of Nε–CBZ-L-lysine **7** to CBZ-L-oxylysine **5**.

Carbobenzoxy(CBZ)-L-oxylysine **5** is an intermediate needed for synthesis of ceronapril (29), a ACE inhibitor **6** being developed for treatment of hypertension (Scheme 3). The precursor of this intermediate, L-oxylysine, has also been used to prepare an analog of the antibiotic, butirosin (30). CBZ-L-oxylysine **5** has been prepared from L-lysine by diazotization with sodium nitrite and sulfuric acid followed by treatment of the product L-oxylysine with benzyl chloroformate (29) As an alternative approach, we have explored the enzymic oxidation of L-lysine or CBZ-L-lysine **7** to the keto acid **8** followed by reduction using a dehydrogenase with appropriate enantioselectivity.

The enzymatic approach to synthesis of CBZ-L-oxylysine **5** is shown in Scheme 3. *C. adamanteus* (rattlesnake) venom L-amino acid oxidase has been reported to oxidize CBZ-L-lysine **7** to the corresponding keto acid **8** (31) and L-HIC dehydrogenase from *Lactobacillus confusus* was found to convert the keto acid **8** to CBZ-L-oxylysine **5** (32). As snake venom enzyme is highly expensive, L-lysyl 2-oxidase from *Trichoderma viride* was tested as a microbial alternative to the snake venom enzyme. Substrate specificity studies (Table 7) showed that the activity of the enzyme is greatly reduced with ε- substituted lysine derivatives. Activity with Nε-formyl-, acetyl-, trifluoroacetyl-, t-butoxycarbonyl(Boc)-, or CBZ-L-lysine was less than 2% of the activity with L-lysine. There was no activity with L-oxylysine. On the other hand, the snake venom oxidase was active with the ε-substituted lysine derivatives but had very little activity with L-lysine. The methyl and ethyl esters of lysine were substrates for the *T. viride* oxidase but were not utilized by the snake venom enzyme.

The product from lysine oxidation by *T. viride* lysyl oxidase was not a substrate for L-HIC dehydrogenase (Table 7) or for several other dehydrogenases that were screened at pH 7.4 or 9. With snake venom L-amino acid oxidase, Nε-formyl-, acetyl-, CBZ-, or trifluoroacetyl-L-lysine were good substrates when the oxidation was coupled to L-HIC dehydrogenase, except, Nε-t-Boc-L-lysine was not an effective substrate. The product of lysine oxidation by the *T. viride* enzyme has been shown to cyclize to a Schiff base (33) and this may interfere with utilization by L-HIC dehydrogenase. Alternatively, although L-HIC dehydrogenase shows rather broad specificity (34) it may require a substituent on the ε-amino of lysine for activity.

Synthesis of CBZ-L-oxylysine 5

T. viride lysyl oxidase (0.3 u/ml) coupled to L-HIC (0.8 u/ml) dehydrogenase was able to convert 1 g/L CBZ-L-lysine **7** to 1 g/L CBZ-L-oxylysine **5** after 48 h in a reaction mixture that also contained 1 mM NAD, 0.2 M sodium formate, 0.7 u/ml formate dehydrogenase, and 1250 u/ml catalase in 0.1 M potassium phosphate at pH 8. In an effort to find a more active CBZ-L-lysine oxidase activity several microbial strains known to possess L-amino acid oxidase were screened. Four strains of *Proteus*

Table VI. Preparative Scale Reduction of N-(4-(1-oxo-2-chloroacetyl-ethyl) phenyl methane sulfonamide 3 by H. polymorpha

Reaction Time	Yield of alcohol 4 (g/L)	EE of alcohol 4 (%)
4	1.1	ND
8	1.8	ND
12	2.4	ND
16	3.2	99.4
20	3.8	99.5

Bioreduction of 3 to 4 was carried out in a 5-L fermentor. Cell-suspensions (3L) was supplemented with 12 g of substrate and 100 g of glucose. Reaction was conducted at 28°C, 200 RPM. Substrate 3 and product 4 concentrations were determined by GC and the optical purity of 4 was determined by chiral HPLC.

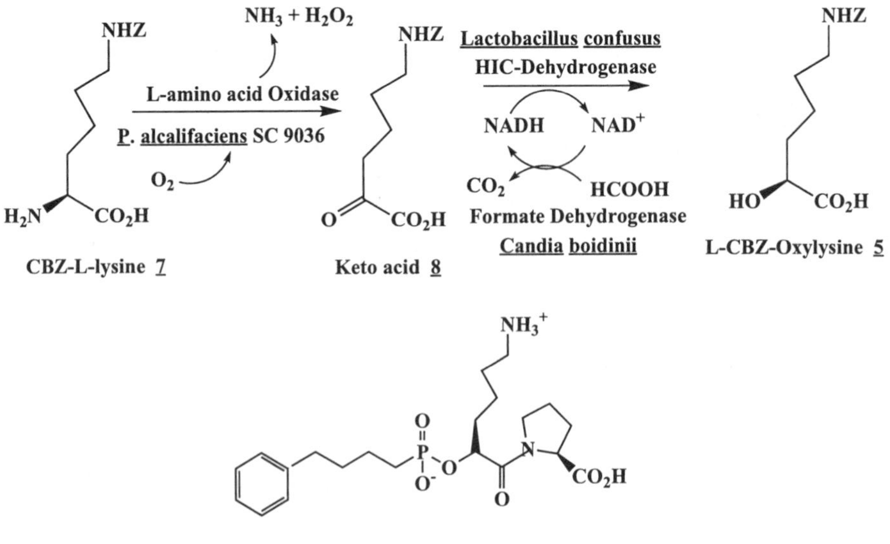

ANTIHYPERTENSIVE DRUG (CERANOPRIL 6)

SCHEME 3

Table VII. Substrate Specificities of L-amino acid Oxidase and L-2-Hydroxyisocaproate (HIC) Dehydrogenase

Enzyme	Substrate	Specific Activity (U/mg Oxidase)
Lysyl Oxidase T. viride	L-Lysine	1.663
	L-Oxylysine	0
	Nε-Acetyl-L-Lysine	0.003
	Nε-Formyl -L-Lysine	0.033
	Nε-BOC-L-Lysine	0
	Nε-Trifluoacetyl-L-Lysine	0.027
	Nε-CBZ-L-Lysine	0.003
	Nε-CBZ-L-Lysine Methyl Ester	0
	L-Lysine Methyl Ester	0.847
	L-Lysine Ethyl Ester	1.083
L-Amino acid Oxidase C. adamanteus	L-Lysine	0
	L-Oxylysine	0.001
	Nε-Acetyl-L-Lysine	0.405
	Nε-Formyl -L-Lysine	0.338
	Nε-BOC-L-Lysine	0.254
	Nε-Trifluoacetyl-L-Lysine	0.406
	Nε-CBZ-L-Lysine	0.405
	Nε-CBZ-L-Lysine Methyl Ester	0.006
	L-Lysine Methyl Ester	0.001
	L-Lysine Ethyl Ester	0
T. viride Lysyl Oxidase coupled to HIC dehydrogenase	L-Lysine	0
	L-Lysine Methyl Ester	0
	L-Lysine Ethyl Ester	0
C. adamanteus L-Amino acid Oxidase coupled to HIC dehydrogenase	Nε-Acetyl-L-Lysine	0.062
	Nε-Formyl -L-Lysine	0.12
	Nε-BOC-L-Lysine	0
	Nε-Trifluoacetyl-L-Lysine	0.028
	Nε-CBZ-L-Lysine	0.12

and *Providencia* were found to oxidize CBZ-L lysine **7** to the keto acid **8**. *P. alcalifaciens* SC9036 gave the highest conversion to keto acid **8** and was selected for further study.

CBZ-L-lysine **7** was converted to CBZ-L-oxylysine **5** in 95% yield by the following procedure. 5.6 g CBZ-L-lysine **7** was added to 1 L of solution containing 0.1 M potassium phosphate, pH 7.4, and 10 g of *P. alcalifaciens* SC9036 cells, and the mixture was incubated at 200 RPM for 27 h at 30°C. CBZ-L-lysine **7** is nearly insoluble at pH 7.4 but went into solution as the reaction proceeded. After 24 h, HPLC analysis indicated complete conversion to the keto acid **8**. After cells were removed by centrifugation, 0.2 M sodium formate, 1 mM NAD, 32 u formate dehydrogenase, and 66 u L-HIC dehydrogenase were added and the solution was incubated for 64 h at 28°. HPLC analysis indicated conversion to 18.9 mM CBZ-L-oxylysine **5**. The isolated product (m.p. 74-75°, standard m.p. 76°) had 98.4% HPLC homogeneity and 98% ee. ^{1}H and ^{13}C-nuclear magnetic resonance(NMR) spectra were identical with those from the chemically produced compound.

L-amino acid oxidase has been reported to be associated with a particulate fraction of the cell in several *Proteus* species. (35,36) Cells, extracts, or the particulate fraction from *P. alcalifaciens* were effective in oxidizing CBZ-L-lysine **7** to the keto acid **8**, but the 101,000 x g supernatant had no activity. This indicates that L-amino acid oxidase from *P. alcalifaciens* is a membrane associated protein.

Antiinfective Drug(Tigemonam)
Enzymatic Synthesis L-β-hydroxyvaline **9** from α-Keto-β-Hydroxyisovalerate **16**.

During the past several years syntheses of α-amino acids have been pursued intensely because of their importance as building blocks of compounds of medicinal interest (37-40). New methods have been developed for the asymmetric synthesis of β-hydroxy-α-amino acids (41-44) because of their utility as starting materials for the total synthesis of monobactam antibiotics. L-β-Hydroxyvaline **9** is a key chiral intermediate needed for the synthesis of tigemonam **10** (Scheme 4), a new orally active monobactam (45-47).

Procedures for the synthesis and resolution of racemic β-hydroxyvaline (48-51) and the asymmetric synthesis of D-β-hydroxyvaline have been reported (52), but the direct synthesis of L-β-hydroxyvaline has not been accomplished. In this section we are describing the direct synthesis of L-β-hydroxyvaline by the reductive amination of α-keto-β-hydroxyisovalerate using leucine dehydrogenase from *Bacillus sphaericus* ATCC 4525. Leucine dehydrogenase from *B. sphaericus* (53), *B. cereus* (54), and *B. megaterium* (55) have been used for the synthesis of branched chain amino acids but not for hydroxylated amino acids. The *B. sphaericus* enzyme has been reported to be inactive for oxidative deamination of serine and threonine (56).

Distribution of α-keto-β-hydroxyisovalerate **16** amination activity in *Bacillus* strains was examined. Leucine dehydrogenase activity is mainly found in *Bacillus* strains. Screening of *Bacillus* strains was carried out to find strain with high specific activity for reductive amination of α-keto-β-hydroxyisovalerate **16** and to identify any

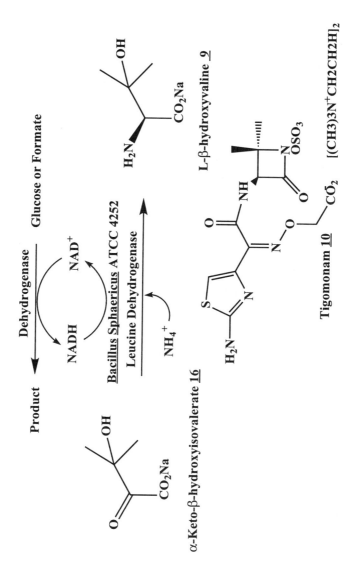

SCHEME 4

strains able to regenerate NADH for the reaction. Of various strains screened (Table 8), *B. sphaericus* ATCC 4525 has the highest specific activity. *B. megaterium* ATCC 38118 has been selected for high levels of glucose dehydrogenase and was able to produce L-β-hydroxyvaline **9** when glucose, NAD (regenerated NADH required), NH_4Cl, and α-keto–β-hydroxy isovalerate **16** were added to cell extracts. All other strains has less than 6% of the glucose dehydrogenase activity of this strain (data not shown), but extracts of a few of the strains were also able to produce L-β-hydroxyvaline **9** when provided with glucose, NAD, NH4Cl, and α-keto-β-hydroxyisovalerate **16**. *B. cereus* ATCC 14579 was the only strain able to produce significant amounts of L-β-hydroxyvaline **9** when intact cells were provided with glucose and α-keto-β-hydroxyisovalerate **16**, giving a maximum conversion of about 50%.

Enzymatic reductive amination of α-keto-β-hydroxyisovalerate **16** by leucine dehydrogenase coupled to formate dehydrogenase was carried out as outlined in Scheme 4. A solution containing 0.5 M α-keto-β-hydroxyisovalerate **16** from ethyl α-keto-β-bromoisovalerate **12**, 2 M ammonium formate, 2mM NAD, 1.5 mg (50 units) leucine dehyodrogenase from *Bacillus* sp., and 40 mg (80 units) formate dehydrogenase from *C. boidinii* in 16 mL of buffer (pH 8) was incubated at 30°C for 41 hours. HPLC analysis of the reaction mixture indicated a 71% conversion of the ester **12** to L-β-hydroxyvaline **9**. L-β-hydroxtvaline **9** was isolated from the reaction. HPLC analysis of the recovered material showed that leucine dehydrogenase produced exclusively the L-isomer of β-hydroxyvaline. α-keto-β-hydroxyisovalerate **16** generated from α-keto-β-bromoisovalerate **11** was also converted to L-β-hydroxyvaline **9** in 82% reaction yield by a similar procedure.

Reductive amination of α-keto-β-hydroxyisovalerate **16** by leucine dehydrogenase coupled to glucose dehydrogenase was carried out. Both enzymes had optimum pH for activity of 8.5. Rates of L-β-hydroxyvaline **9** formation in a coupled system containing both enzymes were similar with 0.5, 1, or 2 mM NAD (Data not shown). When reactions were carried out in a pH stat, solution of α-keto-β-hydroxy isovalerate **16** prepared from 0.5 M **12**, 0.25 M **13**, 0.25 M **14**, 0.25 M **15** [Scheme 5] were converted to L-β-hydroxyvaline **9** in yields of 75, 100, 71 and 83%, respectively.

Vasopeptidase Inhibitor: Antihypertensive Drug
Enzymatic Synthesis of L-6-Hydroxynorleucine **17** from 2-Keto-6-Hydroxyhexanoic Acid **18**

L-6-Hydroxynorleucine **17** is a chiral intermediate useful for the synthesis of a vasopeptidase inhibitor now in clinical trial and for the synthesis of C-7 substituted azepinones as potential intermediates for other antihypertensive metalloprotease inhibitors (57, 58). It has also been used for the synthesis of siderophores, indospicines and peptide hormone analogs (59-65). Previous synthetically useful methods for obtaining this intermediate have involved synthesis of the racemic compound followed by enzymatic resolution. D-amino acid oxidase has been used to convert the D-amino acid to the ketoacid leaving the

Table VIII. Synthesis of L-β-Hydroxyvaline by Bacillus Strains

Strain	Specific Activity (a) (units/mg Protein)	L-β-Hydroxyvaline(mM) (b)	(c)
B. subtilis SC 13794	0.1289 (d)	38	73.2
B. subtilis SC 10253	0.0343	11	69.8
B. megaterium ATCC 39118	0.1232 (d)	66.4	74.5
B. megaterium SC 6394	0.3123 (d)	61	66.4
B. sphaericus ATCC 4225	0.8705	6.5	57.7
B. cereus SC 10856	0.1892	68.4	76.1
B. pumilis SC 8513	0.2079	2.5	68.4
B. licheniformis SC 12148	0.1039	40.5	73.7
B. thuringiensis SC 2928	0.5259	2.8	73.6
B. brevis SC 3812	0.276	0	34.3
B. coagulans SC 9261	0.0194	0	48.3

(a) Assay contained 10 mM α-keto-β-hydroxyisovalerate, 0.3 mM NADH, 0.75 mM NH₄Cl-NH₄OH, pH 9.5
(b) After incubation with 1 M NH4Cl, 1M glucose, 100 mM α-keto-β-hydroxyisovalerate, and 2 mM NAD
© After incubation with 1 M ammonium formate, 100 mM α-keto-β-hydroxyisovalerate, 40 unis/mL formate dehydrogenase, and 2 mM NAD.
(d) L-β-hydroxyvaline was assayed after 68 hours of reaction.

234

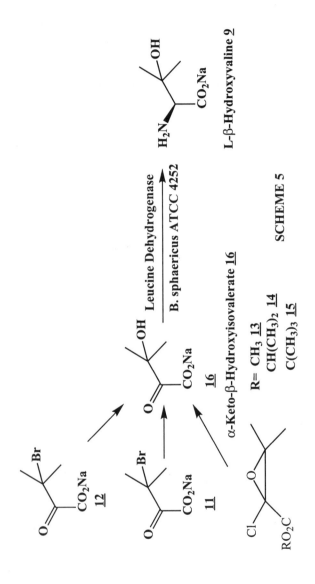

SCHEME 5

L-enantiomer which was isolated by ion exchange chromatography (66). In a second approach, racemic N-acetylhydroxynorleucine has been treated with L-amino acid acylase to give the L-enantiomer (57,64). Both of these resolution methods give a maximum 50% yield and require separation of the desired product from the unreacted enantiomer at the end of the reaction. Reductive amination of ketoacids using amino acid dehydrogenases has shown to be a useful method for synthesis of natural and unnatural amino acids (67-69). We have developed the synthesis and conversion of 2-keto-6-hydroxyhexanoic acid **18** to L-6-hydroxynorleucine **17** by reductive amination using beef liver glutamate dehydrogenase and glucose dehydrogenase from *Bacillus* sp. for regeneration of NADH (70). To avoid the lengthy chemical synthesis of the ketoacid, a second route was developed to prepare the ketoacid by treatment of racemic 6-hydroxy norleucine (readily available from hydrolysis of 5-(4-hydroxybutyl) hydantoin **19**) with D-amino acid oxidase from porcine kidney or *Trigonopsis variabilis* and catalase followed by the reductive amination procedure to convert the mixture to L-6-hydroxynorleucine.

2-Keto-6-hydroxyhexanoic acid **17**, converted completely to L-6-hydroxynorleucine **18** by phenylalanine dehydrogenase from *Sporosarcina* sp. and beef liver glutamate dehydrogenase, with formate dehydrogenase for regeneration of NADH (70). Beef liver glutamate dehydrogenase was used for preparative reactions at 100 g/L substrate concentration. As depicted in Scheme 6, 2-keto-6-hydroxyhexanoic acid **17**, sodium salt, in equilibrium with 2-hydroxy tetrahydropyran-2-carboxylic acid , sodium salt **20**, is converted to L-6-hydroxynorleucine. The reaction requires ammonia and reduced nicotinamide adenine dinucleotide (NADH). Nicotinamide adenine dinucleotide (NAD) produced during the reaction was recycled to NADH by the oxidation of glucose to gluconic acid using glucose dehydrogenase from *Bacillus megaterium*. Reaction was complete in about 3 h with reaction yields of 89-92%, and enantiomeric excess was >98%.

Chemical synthesis and isolation of 2-keto-6-hydroxyhexanoic acid required several steps. In a second more convenient process (Scheme 7), the ketoacid was prepared by treatment of racemic 6-hydroxynorleucine (produced by hydrolysis of 5-(4-hydroxybutyl) hydantoin) with D-amino acid oxidase and catalase. After the optical purity of the remaining L-6-hydroxynorleucine had risen to >99%, the reductive amination procedure was used to convert the mixture containing 2-keto-6-hydroxyhexanoic acid and L-6-hydroxynorleucine entirely to L-6-hydroxynorleucine with yields of 91 to 97% and ee of >98%. Sigma porcine kidney D-amino acid oxidase and beef liver catalase or *Trigonopsis variabilis* whole cells (source of oxidase and catalase) were used successfully for this transformation

Enzymatic Synthesis of Allysine Ethylene Acetal 21

(S)-2-amino-5-(1,3-dioxolan-2-yl)-pentanoic acid (allysine ethylene acetal, **21**) is one of three building blocks used for an alternative synthesis of omapatrilat, a vasopeptidase inhibitor (57,58). It has previously been prepared in an 8-step synthesis from 3,4-dihydro-2H-pyran for conversion into 1-piperideine-6-carboxylic acid, an intermediate for biosynthesis of β-lactam antibiotics (58). Our goal was to prepare compound **21** by a simpler, more convenient route for synthesis of omapatrilat.

236

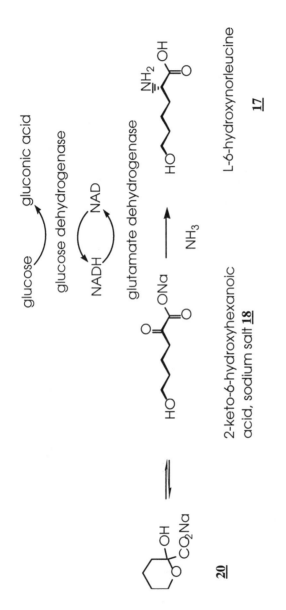

2-keto-6-hydroxyhexanoic
acid, sodium salt **18**

L-6-hydroxynorleucine

17

SCHEME-6

237

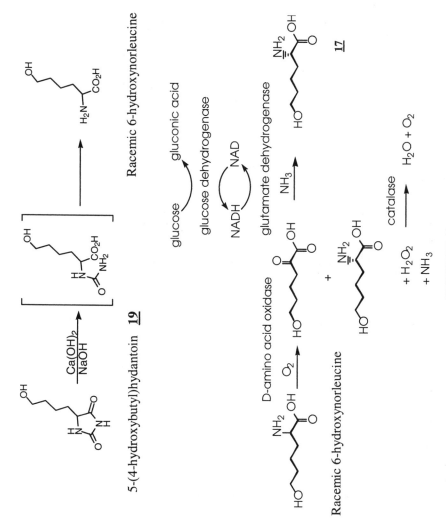

SCHEME 7

The reductive amination of ketoacid acetal **22** to acetal amino acid **21** is carried out using phenylalanine dehydrogenase from *Thermoactinomyces intermedius* [Scheme 8]. The reaction requires ammonia and reduced nicotinamide adenine dinucleotide (NADH). Nicotinamide adenine dinucleotide (NAD) produced during the reaction was recyled to NADH by the oxidation of formate to CO_2 using formate dehydrogenase (71). An initial process is described using heat-dried cells of *Thermoactinomyces intermedius* ATCC 33205, as a source of phenylalanine dehydrogenase, and heat-dried cells of methanol-grown *Candida boidinii* as a souce of formate dehydrogenase. An improved process using phenylalanine dehydrogenase from *Thermoactinomyces intermedius* expressed in *Escherichia coli BL21(DE3)* (pPDH155K) [SC16144] in combination with *Candida boidinii*, and a third generation process using methanol-grown *Pichia pastoris* containing endogenous formate dehydrogenase and expressing *Thermoactinomyces intermedius* phenylalanine dehydrogenase are also described.

Reductive amination of compound 22 by amino acid dehydrogenases

Glutamate, alanine, leucine, and phenylalanine dehydrogenases (listed in order of increasing effectiveness) converted **22** to the desired amino acid **21** (Table 9). The product was identical by HPLC and MS analysis to a chemically synthesized sample. Some alternative sources of phenylalanine dehydrogenase were tested. *Sporosarcina ureae* strains SC 16048 and SC 16049 had respective specific activities of 0.996 and 0.862 u/mg for reductive amination of phenylpyruvate, but amination of of **22** was much slower than with the enzyme from *Thermoactinomyces*. Using an extract of *Thermoactinomyces intermedius* ATCC 33205 as a source of phenylalanine dehydrogenase and Boehringer Mannheim formate dehydrogenase for NADH regeneration increased the the estimated yield was increased to 80%, and the process was developed using this enzyme combination. Heat-dried cells of *Thermoactinomyces intermedius* and *Candida boidinii* SC13822 grown on methanol could be used for the reaction.

Enzyme production by fermentation

Enzyme activities in cells recovered from fermentations and fermentor productivities are shown in Table 10 . *Thermoactinomyces intermedius* gave a useful activity on a small scale (15 L), but lysed soon after the end of the growth period making recovery of activity difficult or impossible on a large scale (4000 L). The problem was solved by cloning and expressing the *Thermoactinomyces intermedius* phenylalanine dehydrogenase in *Escherichia coli* , inducible by isopropyl-thiogalactoside. Fermentation of *Thermoactinomyces intermedius* yielded 184 units of phenylalanine dehydrogenase activity per liter of whole broth in 6 hours. At harvest the fermentor needed to be cooled rapidly, because the activity was unstable. In contrast, *E. coli* produced over 19,000 units per liter of whole broth in about 14 hours and was stable at harvest.

Candida boidinii grown on methanol was a useful source of formate dehydrogenase as has been shown previously (72). In order to recover the cells on a large scale, it was helpful to add 0.5% methanol to stabilize the cells. *Pichia pastoris*

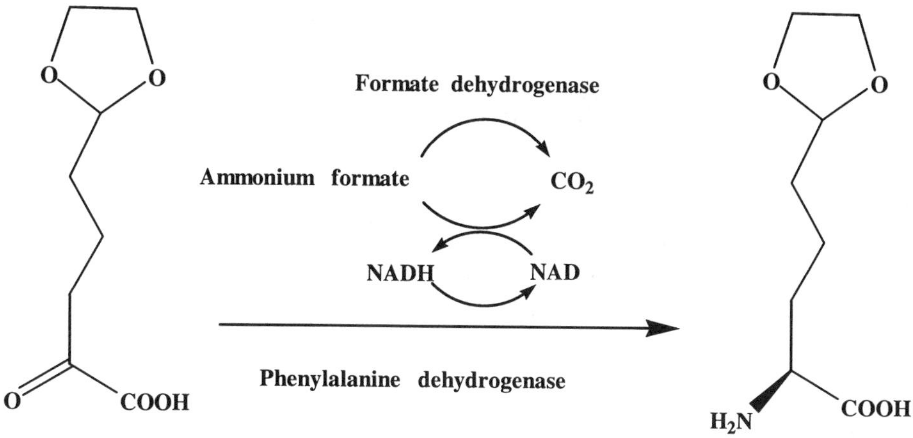

Formate dehydrogenase

Ammonium formate — CO_2

NADH — NAD

Phenylalanine dehydrogenase

22

21

SCHEME 8

Table IX. Reductive amination of ketoacid 18 by amino acid dehydrogenase

Dehydrogenase	Source (1)	Amount (Units)	Amino acid Produced (mM)
Glutamate	Beef liver	76	1.03
Alanine	Bacillus subtilis	35.7	11.77
Leucine	Bacillus sphaericus	22	14.01
Phenylalanine	Sporosarcina sp.	12.6	51.7

(1) All enzymes were obtained from Sigma Chemicals except leucine dehydrogenase was partially purified from B. sphaericus ATCC 4525.

grown on methanol is also a useful source of formate dehydrogenase (73). Expression of *Thermoactinomyces intermedius* phenylalanine dehydrogenase in *Pichia pastoris*, inducible by methanol, allowed both enzymes to be obtained from a single fermentation. The expression of the two activities during a *Pichia pastoris* fermentation is shown in Figure 2. Formate dehydrogenase activity /g wet cells was 2.7-fold greater than for *Candida boidinii* and fermentor productivity was increased by 8.7-fold compared to *Candida boidinii*. Fermentor productivity for phenylalanine dehydrogenase in *Pichia pastoris* was about 28% of the *Escherichia coli* prductivity.

Reductive amination reactions

Formate dehydrogenase has been reported to have a pH optimum of 7.5 to 8.5 (72). The pH optimum for the reductive amination of **22** by an extract of *Thermoactinomyces intermedius* was found to be about 8.7. Reductive amination reaction were carried out at pH 8.0. A summary of lab scale batches is shown in Table 11. The time course for representative batch showing conversion of ketoacid **22** to amino acid **21** is presented in Figure 3 using *Escherichia coli/ Candida boidinni heat-dried cells.*

The procedure using heat-dried cells of *Escherichia. coli* containing cloned phenylalanine dehydrogenase and heat-dried *Candida boidinii* was scaled up (Table 12). A total of 197 kg of **21** was produced in three 1600-L batches using a 5% concentration of substrate **22** with an average yield of 91.1 M % and ee greater than 98%.

Third generation procedure, using dried recombinant *Pichia pastoris* containing *Thermoactinomyces intermedius* phenylalanine dehydrogenase inducible with methanol, and endogenous formate dehydrogenase induced when *Pichia pastoris* is grown in medium containing methanol, allowed both enzymes to be produced during a single fermentation, and they were conveniently produced in about the right ratio that was used for the reaction. The *Pichia* reaction procedure had the following modifications of the *Escherichia coli/ Candida boidinii* procedure: concentration of substrate was increased to 100 g/L, and 1/4 the amount of NAD were used and dithiothreitol was omitted. The procedure with *Pichia pastoris* was also scaled up to produce 15.5 kg of **22** with 97 M % yield and ee greater than 98% (Table 12) in a 180-L batch using 10% ketoacid concentration.

Enzyme Immobilization

For reusability, formate dehydrogenase could be immobilized on Eupergit C and phenylalanine dehydrogenase on Eupergit C250L. The immobilized enzymes were tested for reusability in a jacketed reactor at maintained at 40° C, and were used 5 times for the conversion of **22** to **21** without much loss of any activity and productivity. At the end of each reaction, the solution was drained from the reactor through a 80/400 mesh stainless steel sieve, which retained the immobilized enzymes, then the reactor was recharged with fresh substrate solution. After 5 reuse, the reaction rate was decreased, however, the original reaction rate was restored in the seventh reuse studies by addition of formate dehydrogenase.

Table X. Activities and productivities of phenylalanine dehydrogenase and formate dehydrogenase for various strains grown in fermentor

Enzyme	Strain	Specific activity (U/g wet cells)	Volumatric activity (U/L of broth)	Producivity (U/L/week)
Phenylalanine dehydrogenase	Thermoactinomyces intermedius	510	185	900
	Escherichia coli	10,000	24,000	94,000
	Pichia pastoris	ND	14,500	25,000
Formate dehydrogenase	Candida boidinii	9	120	350
	Pichia pastoris	26	1950	3200

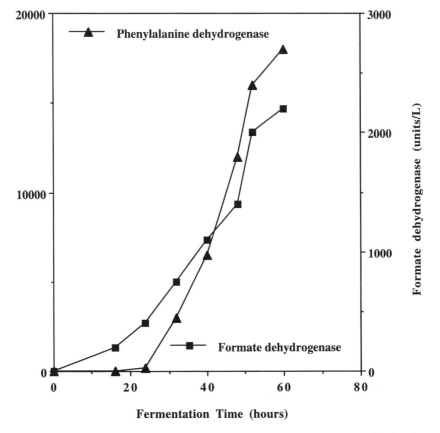

Figure 2: Formate dehydrogenase and phenylalanine dehydrogenase production in *Pichia pastoris* in a 55-L fermentation.

Table XI. Laboratory scale (1 L) batches for reductive amination reactions

Phenylalanine dehydrogenase source	Formate dehydrogenase source	Reaction yield (%)	EE of Product (%)
T. intermedius	Candida boidinii	85	>99
Escherichia coli	Candida boidinii	90	>99
Pichia pastoris	Pichia pastoris	94	>99

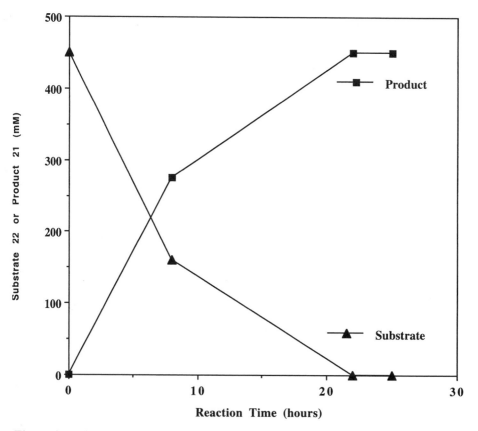

Figure 3: Reductive amination of keto acid **22** to amino acid **21** using heat-dried cells of recombinant *E. coli* (source of phenylalanine dehydrogenase) and *Candida boidinii* (source of formate dehydrogenase).

Table XII. Large scale batches for reductive amination reactions

Phenylalanine dehydrogenase source	Formate dehydrogenase source	Keto acid input (Kg)	Amino acid output (Kg)	Reaction yield (M%)	EE of Amino acid (%)
Escherichia coli	Candida boidinii	80.17	62.4	92	>99
Escherichia coli	Candida boidinii	79.96	66.75	96	>99
Escherichia coli	Candida boidinii	89.6	67.61	86	>99
Pichia pastoris	Pichia pastoris	18.05	15.51	97.5	>99

Thermoactinomyces intermedius IFO14230 (ATCC 33205) was first identified as a source of phenylalanine dehydrogenase by Ohshima *et al.* (74). The enzyme was purified and characterized then cloned and expressed in *Escherichia coli* by the same workers (75). The enzyme was reported to be rather specific for deamination of phenylalanine (76), and to carry out amination of some ketoacids at a much lower rate than amination of phenylpyruvate (74). In our screening, the enzyme was the most effective amino acid dehydrogenase identified for the bioconversion of **22** to **21**. Formate dehydrogenase from *Candida boidinii* was introduced by Shaked and Whitesides (77), and by Kula, Wandrey and coworkers (69) for regeneration of NADH. The advantages of this enzyme reaction are that the product CO_2 is easy to remove, and the negative reduction potential ($E^{\circ\prime}=^-0.42$ v) for the formate dehydrogenase reaction drives the reductive amination to completion.

References

1.   Sih, C. J., and Chen, C. S. *Angew. Chem.Int. Engl.23*, **1984**, 570-578.

2   Jones, J. B. Enzymes in organic synthesis. *Tetrahedron*, **1986**, *42*, 3351-3403.

3.   Csuz, R. and Glanzer, B. I.. *Chem. Rev.***1991**, *91*, 49-97.

4.   Crosby, J. *Tetrahedron. 47*, **1991**, 4789-4846.

5.   Simon, H., Bader, J., Gunther, H., Neumann, S., and Thanos, J. *Angew. Chem. Int. Engl.* **1985**, *24*, 539-553.

6   Stinson, S. C. Chiral drugs. *Chem. Eng. News,* Sept. 28. **1992**, 46-79.

7   Santaneillo, E., Ferraboschi, P., Grisenti, P., and Manzocchi, A. *Chem. Rev.* **1992**, *92*, 1071-1140.

8   Margolin, A. L. *Enzyme. Microb. Technol.* **1993**, *15*, 266-280 (1993).

9.   Wong, C-H and Whitesides, G. M. Tetrahedron Organic Chemistry Series, **1992**, 12. Pergamon, Elsevier Science Ltd. USA. (1994).

10.   Patel, R. N., Banerjee, A., Liu, M., Hanson, R. L., Ko, R. Y., Howell, J. M., and Szarka, L. J. *Biotechnol. Appl. Biochem.* **1993**, *17,* 139-153.

11.   Patel, R. N., Banerjee, A., Mc Namee, C. J., and Szarka, L. *J. Appl. Microbiol Biotechnol.* **1993**, *40*, 241-245.

12.   Hanson, R. L., Singh, J., Kissick, T. P. , Patel, R. N., Szarka, L. J., and Mueller, R. H. *Bioorg. Chem. 18*, **1990**, 116-130.

13.   Hanson, R. L., Bembenek, K. S., Patel, R. N., and Szarka, L. J. *Appl. Microbiol. Biotechnol.*, **1992**, *37*, 599-603.

14.   Ferris, C. D., D. J. Hirsch, B. P. Brooks, and S. H. Snyder. *J. Neurochemistry.*, **1991**, 57, 729-737.

15.   Junien, J. L. and B. E. Leonard. *Clinical Neuropharmacolog.*, **1989**, *12*: 353-374.

16.   Martin, W. R., C. G. Eades, J. A. Thompson, R. E. Huppter, and P. E. Gilbert. J. *Pharmacol. Exp. Ther.* 1976, *197*, 517-539.

17.   Walker, J. M., W. D. Bower, F. O. Walker, R. R. Matsumoto, B. D. Costa, and K. C. Rice. *Pharm. Reviews.* **1992**, *42*, 355-402.

18.   Steinfels, G. F., S. W. Tam, and L. Cook. *Neureopsychopharmacology.* **1989**, 2: 201-207.

19. Massamiri, T., and S. P. Duckles. *J. Pharmacology and Experimental Therapeutics.* **1990**, *253*, 124-129 (1990).

20. Martinez, J. A. and L. Bueno. *European J. Pharmacology.* **1991**, *202*: 379-383.

21. Taylor, D. P., M. S. Eison, S. L. Moon, and F. D. Yocca.. *Adv. Neuropsy. Psychopharm.* **1991**, *1*: 307-315.

22. Yevich, J. P., W. G. Lobeck. **1986**. U.S. Patent # 4605655.

23. Yevich, J. P., J. S. New, D. W. Smith. *J. Med. Chem.* **1986**, *29*: 359-369.

24. Bradford, M. M. *Anal. Biochem.***1976**, *72*, 248-254.

25. Larsen, A. A., and Lish P. M. *Nature.***1964**, *203*- 1283-1284.

26. Uloth, R. H., Kirk, J. R. , Gould, W. A., and Larsen, A. A. *J. Med. Chem.* **1966**, *9,* 88-96.

27. Lish, P.A., Weikel, J. H, and Dungan K. W. *J. Pahrma and Exper. Therapeutics.* **1965**, *149*, 161-173.

28. Somani, P., and Bachand, T. *Eur. J. Pharma,* **1969**, *7*, 239-247.

29. Karanewsky, D.S., Badia, M. C., Cushman, D. W., DeForrest, J. M. Dejneka, T., Loots, M. J., Perri, M. G., Petrillo, E. W. Jr., and Powell, J. R.. *J. Med. Chem.* **1988**, *31*, 204-212.

30. Haskell, T. H., Rodebaugh, R., Plessas, N., Watson, D., and Westland, R. D. *Carbohyd. Res.* **1973**, *28,*263-280.

31. Meister, A. *J. Biol. Chem.* **1954**, *206*, 577-585.

32. Robison, R. S., Doremus, M. G. and Szarka, L. J., unpublished results

33. Kusakabe, H., Kodama, K., Kuninaka, A., Yoshino, H., Misono, H., and Soda, K. *J. Biol. Chem.* **1980**, *255*, 976-981 .

34. Schütte, H., Hummel, W., and Kula, M. R.. *Appl. Microbiol. Biotechnol.* **1984**, *19,*167-176.

35. Pelmont, J., Arlaud, G., and Rossat, A. M. *Biochimie.* **1972**, *54*, 1359-1374.

36. Duerre, J. A. and Chakrabarty, S. *J. Bacteriol.* **1975**, *121*, 656-663.

37. Williams, R. M. In Synthesis of Optically Active a-amino Acids. (Baldwin, J. E. and Magnus, P. D. Eds.) **1989**, vol. 7, Pergamon, Oxford/ New York.

38. Kamphuis, J., Boesten, W. H., Broxterman, Q. B., Hermes, H. F. M., van Balken, J. A. M. Meijer, E. M., and Shoemaker, H. E. *Adv. in Biochem. Engin. Biotechnol.* **1992**, *42*, 134-186.

39. Sykes, R. B., Cimarusti, C. M., Bonner, D. P., Bush, K., Floyd, D. M., Georgopadakou, N. H., Koster, W. H., Liu, W. C., Parker, W. L., Principle, P. A., Rathnum, M. L., Slusarchyk, W. A., Trejo, W. H., and Wells, J. S. *Nature (London) 291*, 489-491.

40. Parker, W. L., O'sullivan, J., and Sykes, R. B. *Adv. Appl. Microbiol.* **1986**, *31*, 181-205.

41. Roemmele, R. C. and Rapoport, H. *J. Org. Chem.* **1989**, *54*, 1866-1875.

42. Ito, Y., Sawamura, M., Shirakawa, E., Hayashizaki, K., and Hayashi, T. *Tetrahedron.* *44*, 5253-5262.

43. Guanti, G., Banfi, L., and Narisano, E. *Tetrahedron.* **1988**, *44*, 5553-5562.

44. Evans, D. A., Sjogren, E. B., Weber, A. E., and Conn, R. E.*Tetrahedron Lett.* **1987**, *28*, 39-42.

45. Gordon, E. M., Ondetti, M. A., Pluscec, J., Cimarusti, C. M., Bonner, D. P., and Sykes, R. B. *J. Amer. Chem. Soc.* **1982**, *104*, 6053-6060.
46. Parker, W. L., Cohen, E. M., and Koster, W. H. **1988**, US Patent 474122.
47. Slusarchyk, W. A., Dejneka, T., Gougoutas, J. Z., Koster, W. H., Kronenthal, D. R., Malley, M. F., Perri, M. G., Routh, F. L., Sundeen, J. E., Weaver, E. R., Zahler, R., Godfrey, J. D., Mueller, R. H., and Langen, D. J. *Tetrahedron Lett.* *27*, 2789-2792.
48. Godfrey, J. D., Mueller, R. H., and Van Langen, D. J. *Tetrahedron Lett.* *27*, 2793-2796.
49. Edwards, G. W. and Minthorn, M. L. Jr. *Canad. J. Biochem.* **1968**, *46*, 1227-1230.
50. Berse, C. and Bessette, P. *Canad. J. Chem.* **1971**, *49*, 2610-2611.
51. Beyerman, H. C., Maat, L., De Rijke, D., and Visser, J. P. *Recl. Trav. Chim. Pays-Bas.* **1967**, *86*, 1057-1060.
52. Schollkopf, U., Nozulak, J., and Groth, U. *Synthesis*, **1982**, 868-870.
53. Wiechmann, R., Wandrey, C., Buckmann, A. F., and Kula, M-R. *Biotechnol. Bioeng.* **1981**, *23*, 2789-2802.
54. Schutte, H., Hummel, W., Tsai, H., and Kula, M. *Appl. Microbiol. Biotechnol.* **1985**, *22*, 306-317.
55. Monot, F., Benoit, Y., Lemal, J., Honorat, A., and Ballerini, D. In Proc. Eur.Cong. Biochem. (Neijssel, O. M., van der Meer, R. R., and Luyben, K. C. A. M. Eds.) **1987**, vol. 2, pp. 42-45, Elsevier, Amsterdam. (1987).
56. Oshima, T., Misono, H., and Soda, K. *J. Biol. Chem.* **1978**, *253*, 5719-5725.
57. Robl, J. A.; Sun, C; Stevenson, J;, Ryono, D. E.; Simpkins, L. M.; Cimarusti, M. P.; Dejneka, T; Slusarchyk, W. A.; Chao, S; Stratton, L.; Misra, R. N. ; Bednarz, M. S.; Asaad, M.M.; Cheung, H. S.; Aboa-Offei, B. E.; Smith, P.L.; Mathers, P. D.; Fox, M.; Schaeffer, T. R.; Seymour, A. A.; Trippodo, N. C. *J. Med. Chem.* 1997, *40*, 1570-1577.
58. Robl, J. A.; Cimarusti, M.P. *Tetrahedron Letters.* **1994**, *35*, 1393-1396.
59. Maurer, P. J.; Miller, M. J. *J. Org. Chem.* **1981**, *46*, 2835-2836.
60. Maurer, P.J.; Miller, M. J. *J. Am. Chem. Soc.* **1982**, *104*, 3096-3101.
61. Maurer, P.J.; Miller, M. J. *J. Am. Chem. Soc.* **1983**, *105*, 240-245.
62. Culvenor, C.C. J.; Foster, M. C.; Hegarty, M.P. *Aust. J. Chem.* **1971**, *24*, 371-375.
63. Bodanszky, M; Martinez, J.; Priestly, G. P.; Gardner, J. D.; Mutt, V. *J. Med. Chem.* **1978**, *21*, 1030-1037.
64. Dreyfuss, P. *J. Med. Chem.* **1974**, *17*, 252-257.
65. Kern, B. A.; Reitz, R. H. *Agric. Biol. Chem.* **1978**, *42*, 1275.
66. Bommarius, A. S. In *Enzyme Catalysis in Organic Synthesis*:; Drauz, K. and Waldmann, H., Ed.; VCH: Weinheim, 1995; Vol II, pp 633-641 and cited references.
67. Galkin, A.; Kulakova, L.; Yoshimura, T.; Soda, K.; Esaki, N. *Appl. Environ. Microbiol.* 1997, *63*, 4651-4655.

68.   Rumbero,A., Martin, J. C., Lumbreras, M. A., Liras, P. and Esmahan, C. *Bioorg. Med. Chem.* 1995, *3*, 1237-1240

69.   Kula, M. R. and Wandrey, C. *Methods in Enzymology* **1987**, *136*, 9-21.

70.   Hanson, R. L., Schwinden, M. D., Banerjee, A., Brzozowski, D. B., Chen, B-C., Patel, B. P., McNamee,C. G., Kodersha, G. A., Kronenthal, D. R., Patel, R. N. and Szarka, L. J. *J. Bioorg. Med. Chem.* **1999**, 7, 2247-2252.

71.   Hanson, R. L.; Howell, J.; LaPorte, T.; Donovan, M-J.; Cazzulino, D.; Zannella, V.; Montana, M.; Nanduri, V.; Scwartz, S.; Eiring R.; Durand, S.; Wasylyk, J.; Parker, L.; Liu, M.; Okuniewicz, F.; Chen, B-J, Harris, J.; Natalie, K.; Ramig, K.; Swaminathan, S.; Rosso, V.; Pack, S.; Lotz, B.; Bernot, P.; Rusowicz, A.; Lust D.; Tse, K.; Venit, J.; Szarka, L.; Patel, R. N. *Enzyme Microbial Technol.*(in press).

72.   Schütte, H. Flossdorf, J. Sahm, H. and Kula, M-R. *Eur. J. Biochem.* **1976**, *62*, 151-160.

73.   Hou, C. T., Patel, R. N., Laskin, A. I. and Barnabe, N. *Arch. Biochem. Biophys.* **1982**, *216*, 296-305

74.   Ohshima, T.; Takada, H.; Yoshimura, T.; Esaki, N.; Soda, K. *J. Bacteriol.* **1991**, *173*, 3943-3948.

75.   Ohshima, T.; Nishida N.; Bakthavatsalam, S.; Kataoka, K.; Takada, H.; Yoshimura, T.; Esaki, N.; Soda, K. *Eur. J. Biochem.* **1994**, *222*, 305-309.

76.   Ohshima, T., Sugimoto, H. and Soda, K. *Anal. Lett* **1988**, *21*, 2205-2215.

77.   Shaked, Z. and Whitesides, G. M. *J. Am. Chem. Soc.* **1980**, *102*, 7104-7105.

Chapter 16

Production of chiral Amines with ω-Transaminase

Jong-Shik Shin and Byung-Gee Kim

School of Chemical Engineering and Institute for Molecular Biology and Genetics, Seoul National University, Seoul, Korea

Various chiral amines were produced using (S)-specific ω-transaminase from *Vibrio fluvialis* JS17 screened from soil microorganisms. The ω-transaminase shows broad substrate specificity and high enantioselectivity. Product and substrate inhibitions were the major obstacles to make the reaction successful. To reduce the inhibitions and to use high concentration of the substrates, three processes for kinetic resolutions, i.e. two-liquid-phase system, enzyme membrane reactor, and packed-bed reactor, were compared. A membrane contactor was used to extract inhibitory ketone for the last two processes. Using the reaction processes, kinetic resolutions of the chiral amines (100~500 mM) were successfully carried out with *ee*>95 % of (R)-amines. Asymmetric synthesis of (S)-amines using prochiral ketone was also attempted and whole cell reaction proved to be successful due to the efficient removal of the product inhibition by pyruvate. However, the asymmetric reaction rate is much lower than the resolution reaction rate.

Chiral amines are of great importance for the pharmaceutical and fine-chemical industries. They are used as intermediates in the synthesis of various drugs, resolving agents for the preparation of optically pure carboxylic acids, and auxiliaries for chiral induction.[1-3] Owing to its great utilities, several biocatalytic methods to prepare the chiral amines have been studied extensively.[2-8] Those methods can be classified into two categories by the enzyme used, hydrolase-catalyzed aminolysis in non-aqueous medium and transamination in aqueous medium. In the case of hydrolase-catalyzed aminolysis[2,4-6], subtilisin and lipase showed high enantioselectivities. Theoretically, it is possible to recover both enantiomers as an amide product and an unreacted amine after the reaction.

However, drastic decrease in enzyme activity in organic medium is a common problem. In the case of kinetic resolution by ω-transaminase[3,7,8], only an unreacted amine enantiomer can be recovered. Therefore, maximum yield is always lower than 50 %. However, full enzyme activity can be utilized because the enzyme reaction is carried out in aqueous medium. In addition, the asymmetric synthesis of chiral amines can be also executed.

Although the ω-transaminase displays broad substrate specificity, very high reaction rate, and good stability, product and substrate inhibitions were the major problems in constructing a successful production system. Therefore, reaction engineering to alleviate the inhibitions has been studied and successful methods to overcome the problems in both kinetic resolution and asymmetric synthesis were devised.

In this article, we present the properties of ω-transaminase reaction and its application to produce chiral amines via both kinetic resolution and asymmetric synthesis.

Properties of ω-Transaminase Reaction

We screened and isolated many microorganisms showing ω-transaminase activity toward various (S)-amines by enrichment culture method. Among them, *V. fluvialis* JS17 showed the highest specific activity and further studies were carried out with the strain.[14]

Mechanism of ω-Transamination

Transaminase[12] is a pyridoxal 5′-phosphate (PLP)-dependent enzyme. The coenzyme acts as an acceptor and donor of the amino group, shuttling between the aldehydic (PLP) and amino (pyridoxamine phosphate, PMP) forms. Unlike α-transaminase reaction, ω-transaminase can transfer the amino group from secondary amine compounds not bearing carboxyl group to carbonyl group and *vice versa*. Therefore, amino donor and acceptor substrates in the ω-transamination are chemically different from the corresponding aminated and deaminated product, respectively. For example, when (S)-α-methylbenzylamine ((S)-α-MBA) is deaminated with pyruvate as shown in Figure 1, amine and keto acid are substrates whereas amino acid and ketone are products. In contrast, substrates and products have identical chemical properties in α-transaminase reaction, i.e., the amino donor is amino acid and the amino acceptor is keto acid in both forward and reverse reactions. Therefore, equilibrium constant of the reactions that α-transaminases involve is around unity, which makes the reaction yield low. However, the equilibrium constant of the reaction by the ω-transaminase in Figure 1 is high enough not to limit the reaction.[7,8] This fact is the most important difference between the reactions by the two enzymes. Conversely, in the case of asymmetric synthesis starting from achiral ketone and amino donor, the equilibrium constant of the reaction is very low.[13]

Substrate Specificity and Stereospecificity

The ω-transaminase from *V. fluvialis* JS17 shows broad substrate specificity towards both amino donors and acceptors, which is very useful for commercial purposes. After extensive study based on substrate structure-activity relationship for both amino donors and acceptors, we delineated active-site structure of the enzyme with a two-site model consisting of large (L) and small (S) binding pockets as shown in Figure 2. This model is quite useful in predicting substrate specificity as well as stereospecificity. We found that the sizes of L and S binding pockets are comparable to hexyl and ethyl group, respectively (unpublished data).

Amino Donor Specificity

Figure 3 shows the structures of model chiral amines. In the case of amino donor, stereospecificity seems to be determined by different sizes of R_1 and R_2. For example, aromatic group in α-MBA cannot enter S site due to size exclusion and should occupy L site. In result, methyl group occupies S site. Under such conditions, amino group of (S)-enantiomer only can access to internal aldimine plane of the enzyme and undergo further reaction to yield E-PMP and deaminated ketone, i.e. acetophenone. Enantioselectivity of the ω-transaminase is very high and the E value for α-MBA is above 140. If R_1 or R_2 is bulkier than the size of each binding site such as 1-(1-naphthyl)ethylamine and 3-amino-3-phenylpropionic acid, reactivities of the amino donor decrease greatly.

Amino Acceptor Specificity

In the case of amino acceptor, chirality is generated after amination. It seems that carboxyl group adjacent to carbonyl carbon of amino acceptor should be positioned at L site for the transfer of the amino group from E-PMP. Therefore, carboxyl and methyl groups of pyruvate should occupy L and S site, respectively. And then transamination produces L-alanine and E-PLP. Negative charge at the carboxyl group of the amino acceptor does not appear to be essential for the catalysis because esters of keto acids are even more reactive than the keto acids, and aldehydes also show good reactivities (unpublished data).

Substrate Inhibition

The ω-transaminase from *V. fluvialis* JS17 shows inhibitions by both substrates in the forward reaction depicted in Figure 1, whereas no substrate inhibitions were observed in the reverse reaction. The substrate inhibition by pyruvate is much more severe than that by (S)-α-MBA as shown in Figure 4A. The reaction rate begins to fall only at 10 mM of pyruvate. This strong substrate inhibition was also observed with other reactive amino acceptors such as propionaldehyde, butyraldehyde, and 2-ketobutyrate.

The degree of the substrate inhibition by the amino donors was found to be related to the amino donor reactivity. The higher reaction rate of the amino donor

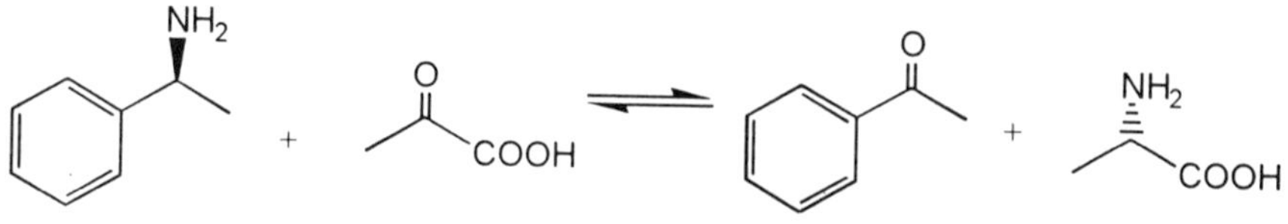

NH2
O
COOH
O
NH2
COOH

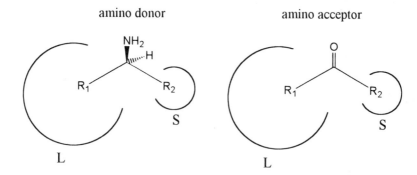

Figure 1. Reaction scheme of ω-transamination.

amino donor
amino acceptor
NH2
H
R1
R2
S
L
O
R1
R2
S
L

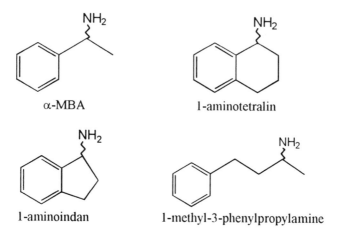

Figure 2. Two-site model of active site of ω-transaminase from V. fluvialis JS17.

NH2
α-MBA
NH2
1-aminotetralin
NH2
1-aminoindan
NH2
1-methyl-3-phenylpropylamine

Figure 3. Structures of model chiral amines.

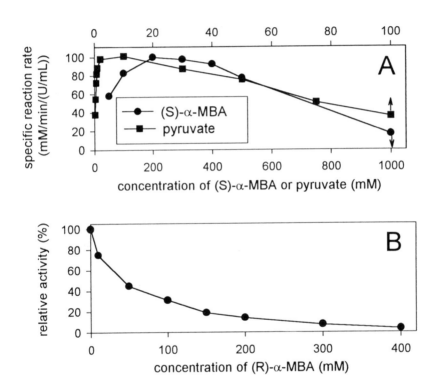

Figure 4. Substrate inhibition. A: by (S)-α-MBA and pyruvate, B: by (R)-α-MBA.

was displayed, the more severe substrate inhibition was observed. For example, in case of (S)-1-aminoindan, which is much more reactive amino donor than (S)-α-MBA, the reaction rate begins to decrease only at 10~30 mM (data not shown). However, when (S)-α-MBA is used as an amino donor, the reaction rate begins to fall from 200 mM as shown in Figure 4A.

The enzyme also shows a substrate inhibition by non-reacting (R)-amine. Figure 4B clearly shows that (R)-α-MBA severely inhibits ω-transamination between (S)-α-MBA and pyruvate although no reaction proceeds between (R)-α-MBA and pyruvate at measurable rate due to the high enantioselectivity. It implies that the inhibition mode by (R)-enantiomer is different from that by (S)-enantiomer. Kinetic study revealed that (R)-enantiomer inhibits the enzyme activity by the formation of Michaelis complex with E-PLP, whereas (S)-enantiomer with E-PMP. This enantioselective substrate inhibition, i.e. different mode of inhibition by each substrate enantiomer, was also observed with other chiral amines (unpublished data).

Product Inhibition

Unlike the substrate inhibition, product inhibition occurs due to the reversibility of the reaction. In the case of the forward reaction in Figure 1 (transamination between (S)-α-MBA and pyruvate), acetophenone exerts more severe product inhibition than L-alanine, suggesting higher binding affinity of acetophenone to E-PMP than that of L-alanine to E-PLP. In the case of the reverse reaction (transamination between acetophenone and L-alanine), the product inhibitions are much more severe than those in the forward reaction, and pyruvate exerts higher inhibition than (S)-α-MBA. In addition to the aforementioned substrate inhibitions, these product inhibitions make ω-transaminase process more complicated.

Reaction Engineering

Although the ω-transaminase shows high reactivity, enantioselectivity, and stability, several problems such as substrate and product inhibitions should be overcome to make the enzyme reaction an attractive method to produce the enantiopure amines.

Factors Affecting ω-Transaminase Reaction

There are several important factors in ω-transamination. First of all, high enantioselectivity is important for obtaining enantiopure amines. All the ω-transaminases we have isolated show sufficiently high (S)-enantioselectivity with $E > 100$, which is high enough to obtain a high purity of chiral amines. Due to the reversibility of ω-transamination, both kinetic resolution and asymmetric synthesis are possible using racemic amine and achiral ketone, respectively. In general, if enantioselectivity of the enzyme is high, asymmetric synthesis is usually favored over kinetic resolution because theoretical maximum yield is 100 %.[11] However, in our case, extremely severe product inhibition and much lower reaction rate made the asymmetric synthesis difficult to be carried out.[13] Therefore, this enzyme appears to be more suitable for kinetic resolution to produce (R)-amines rather than asymmetric synthesis of (S)-amines.

Substrate and product inhibitions are the most critical factors in succeeding the ω-transaminase reaction at high concentrations of amines. Due to the substrate inhibitions by both amine enantiomers and amino acceptor, fed-batch reaction is favored for kinetic resolution over batch reaction. During the kinetic resolution, the product inhibition by ketone can be reduced by incorporating extraction process. In the case of the asymmetric synthesis, high concentrations of the substrates, i.e. achiral ketone and amino donor, are preferable for achieving high reaction rate due to the absence of the substrate inhibition. However, at least one of the products should be removed for reducing the product inhibition as well as shifting the equilibrium.

Sensitivity Analysis of Kinetic Parameters

Kinetic model of the ω-transaminase from *Bacillus thuringiensis* JS64 was developed in the previous study based on ping-pong bi-bi mechanism using King-Altman method.[8] In sensitivity analysis [9], reaction rate is regarded as a function of kinetic parameters as well as time and concentrations of substrates and products.

The sensitivities of the kinetic resolution reaction rate for the Michaelis constants of substrates (i.e. (S)-α-MBA and pyruvate) and products (i.e. acetophenone and L-alanine) are negligible, compared with those for product inhibition constants. The reaction rate shows the highest sensitivity to the changes in the product inhibition constant of (S)-α-MBA (K_{imp}) irrespective of reaction time. This result can give some insights into the kinetics of the ω-transaminase. High equilibrium constant of the forward reaction in Figure 1 requires lower product inhibition constants of (S)-α-MBA and pyruvate (K_{ipp}) than those of acetophenone and L-alanine to meet Haldane relationship. The lower values of K_{imp} and K_{ipp} make the forward reaction proceed as if it is irreversible. The product inhibition by (S)-α-MBA seems to be more responsible for the suppression of the reverse reaction than that by pyruvate, because K_{ipp} is three times larger than K_{imp}. Therefore, even the slight change in K_{imp} would significantly destroy the pseudo-irreversibility of the forward reaction, in result would change net reaction rate to a great extent.

Optimization of the ω-Transaminase Reaction by Simulation

Kinetic model of the ω-transaminase from *V. fluvialis* JS17 was also developed including substrate inhibition by pyruvate and both enantiomers of α-MBA. The kinetic model enables various kinds of optimizations that are difficult and cumbersome to be carried out experimentally. In this article, the effect of substrate inhibition by (R)-amine on the kinetic resolution of racemic amine is analyzed.

Figure 5 shows the comparison of the simulation results between with and without substrate inhibition term for (R)-α-MBA in kinetic rate equation. The substrate inhibition by (R)-enantiomer decreases the reaction rate considerably although only 25 mM of (R)-α-MBA is present during the reaction. This result suggests that the high concentration of the racemic amine substrate is not favored at all for achieving high reaction rate due to the substrate inhibition by (R)-enantiomer in addition to the inhibition by (S)-enantiomer. However, it is not easy to predict the optimal concentration of the amine substrate for the kinetic resolution just by accounting kinetic parameters such as Michaelis constant and inhibition constant.

From the industrial viewpoint, productivity would be one of the most important factors in optimizing the process parameters. We carried out simulations to evaluate the productivity of the kinetic resolution reaction with respect to the initial amine concentrations. In order to eliminate other factors affecting reaction rate from the simulation results, the concentration of pyruvate was set to be constant at 30 mM and ketone product was assumed to be completely extracted. This assumption can be attained by dosing pyruvate continuously and using an efficient extraction system such as membrane contactor.

Figure 6 shows that optimal initial concentration of racemic α-MBA is approximately 110 mM at the given reaction conditions. The productivity begins to decrease gradually above 110 mM, which is caused by the increase in the substrate inhibitions by both enantiomers of α-MBA. This example illustrates that the kinetic resolution process can be easily optimized by the simulation based on the kinetic model.

Kinetic Resolution

After kinetic resolutions of the racemic amines, unreacted (R)-amines can be recovered. Here, we discuss removal method of the ketone product using membrane contactor and compare three reaction processes studied so far.

Removal of Inhibitory Ketone Product

As pointed out previously, the major problem in the kinetic resolution is the inhibition by ketone product and by both substrates. The substrate inhibitions can be overcome by controlling the substrate concentrations at low values. Optimal amino donor concentration can be determined by the simulations based on productivity as shown previously. In the case of amino acceptor, the concentration should be controlled at optimal value by continuous feeding during the operation.

An easy way to overcome the product inhibition is an extractive bioconversion owing to hydrophobicity and high boiling point of the ketone product. In the previous report, it was demonstrated that a two-liquid-phase reaction system was successful in reducing the enzyme inhibition caused by the ketone product.[7] However, interfacial enzyme inactivation was a serious problem in the emulsion state, and constructing a continuous process was difficult due to the troublesome separation of the two liquid phases. To overcome such problems, a membrane

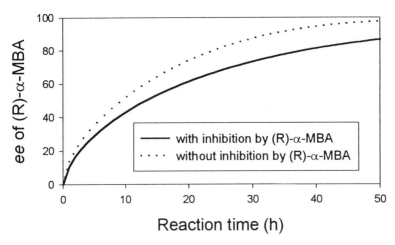

Figure 5. Simulations to predict the effect of substrate inhibition by (R)-α-MBA on the kinetic resolution of α-MBA. Simulation conditions: 50 mM racemic α-MBA, 30 mM pyruvate, and 1 U/mL of enzyme.

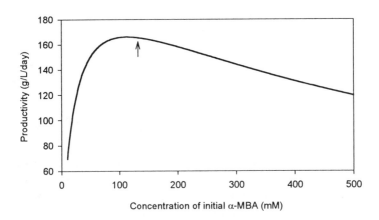

Figure 6. Simulation of kinetic resolution of α-MBA in batch reactor with complete extraction of acetophenone and fixing the pyruvate concentration at 30 mM to optimize racemic amine concentration. Enzyme concentration is 10 U/mL.

contactor (Liqui-Cel® Extra-Flow 2.5 × 8; Hoechst) was used for continuous extraction, which enables selective extraction of ketone from the reaction media.

As shown in Figure 7, the outlet concentration of acetophenone in the organic flow reaches steady state after 30 min, which corresponds to one pass of the aqueous flow through the tube side. The outlet concentrations of acetophenone in the aqueous flow were less than 0.01 mM, suggesting nearly complete extraction. All the partition coefficients of α-MBA, pyruvate, and L-alanine between isooctane and the reaction buffer are nearly zero due to their charge at neutral pH. Therefore, only acetophenone can be extracted selectively from the aqueous solution using the membrane contactor. The extraction method using the membrane contactor was employed in EMR and PBR processes.

Reaction Processes

Two-liquid-phase Reaction System

In the two-liquid-phase reaction system, a water-immiscible solvent was introduced to extract the inhibitory ketone produced in the aqueous reaction medium. Kinetic resolution of α-MBA up to 500 mM was successfully carried out using ω-transaminase from *B. thuringiensis* JS64.[7] The organic phase directly contacts the aqueous phase and serves as both an extractant of inhibitory ketone and a reservoir of amine substrate. The reaction set-up is easy to construct for batch reaction. However, interfacial enzyme inactivation was a serious problem in the emulsion state, and constructing a continuous process was difficult due to the troublesome separation of the two phases, which makes it difficult to reuse the enzyme repetitively.

Enzyme Membrane Reactor

The EMR process enables us to overcome the problems in the biphasic reaction system, such as interfacial enzyme inactivation and difficulty in reusing the enzymes. Flow scheme of the reaction process is shown in Figure 8. In this process, maintenance of the extraction efficiency of the ketone product requires a simple exchange of solvent in the organic reservoir. In addition, pH control is not needed in this set-up because only acetophenone is transferred through the membrane contactor. Therefore, the EMR process is more feasible for scale-up than the two-liquid-phase reaction process. The only problem in the kinetic resolution with the EMR process would be that a highly purified enzyme should be used to achieve the high aqueous flow rates so that the product inhibition can be minimized.

Small-scale kinetic resolutions of α-MBA and 1-aminotetralin with a flat membrane reactor (10 mL working volume) as well as preparative-scale resolutions of 1-aminotetralin using a hollow-fiber membrane reactor (39 mL working volume) were carried out. In the case of the preparative kinetic resolution of 1-aminotetralin (200 mM, 1 L of total substrate reservoir), *ee* of (R)-1-aminotetralin in the substrate reservoir was 95.6 % after 133 h using 390 U of enzyme.

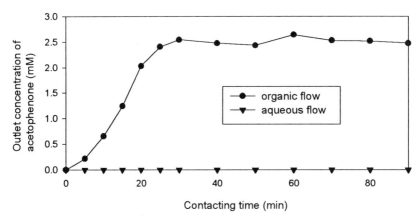

Figure 7. Extraction of acetophenone from the aqueous flow in tube side through the membrane contactor. Flow rates of the aqueous and the organic flow (isooctane) were 300 and 1200 mL/h, respectively. Inlet concentration of acetophenone in the aqueous flow was 10 mM.

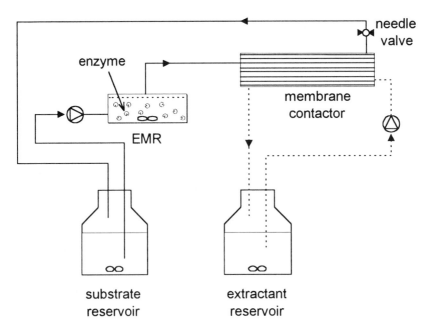

Figure 8. Flow scheme of the kinetic resolution process using flat membrane reactor.

Packed-bed Reactor

The EMR process discussed in the above section has many advantages over the process using immobilized enzymes, such as permitting homogeneous catalysis, no loss of enzyme activity during immobilization, easiness of sterilization, and easy supply of fresh enzyme. However, higher equipment costs and difficulty in operating the plug-flow reaction would be drawbacks.[10] Unlike the EMR, the concentration of inhibitory ketone product builds up with respect to the distance from flow inlet in the reactor. Therefore, to reduce the product inhibition packed-bed reactor (PBR) would be preferable to EMR, because EMR is similar to CSTR and hence the concentration of the inhibitory ketone product in the reactor is the same as the outlet concentration.

We used entrapped whole cells of *V. fluvialis* JS17 in alginate bead for the kinetic resolution with PBR (Figure 9). Immobilizing matrix affects the partitioning of substrates and products, consequently the degree of enzyme inhibitions. Due to the hydrophilicity of alginate matrix, inhibitions by the hydrophilic compounds such as pyruvate and L-alanine increase, whereas those by α-MBA and acetophenone are reduced. Kinetic resolutions of α-MBA, 1-aminoindan, and 1-aminotetralin were carried out successfully using the PBR process (data not shown).

Comparison of the Reaction Processes

Three kinetic resolution processes studied so far have its own pros and cons. In the two-liquid-phase system, any kinds of enzyme form can be used. However, in EMR and PBR process, free and immobilized enzyme can be usually used, respectively. In the two-liquid-phase system, the enzyme inactivation is fast and reuse of the enzyme is difficult but the equipment cost is low and scale-up is easy. In EMR, no mass transfer limitation and equivalent amino acceptor consumption to (S)-amines are advantageous over PBR. In addition, reuse of the enzyme and scale-up are easy. In PBR, reuse of enzyme is easy but the preparation of immobilized whole cell is costly and tedious as long as the immobilized cell is not stable enough for the repetitive use. Availability of the enzyme is important in choosing the type of bioreactor. If purified enzyme is available at low price, despite high equipment cost, the enzyme membrane reactor seems to be a better choice than the packed-bed reactor due to no mass transfer limitation and no loss of enzyme activity during immobilization.

Asymmetric Synthesis

Asymmetric synthesis of chiral compounds using prochiral ketones is usually favored than kinetic resolution using racemic mixture because theoretical yield can reach 100 % and achiral precursors are usually cheaper than racemic mixtures.[11] The ω-transaminase from *V. fluvialis* JS17 has sufficiently high enantioselectivity permitting the production of chiral amines of very high enantiomeric purity via asymmetric synthesis.

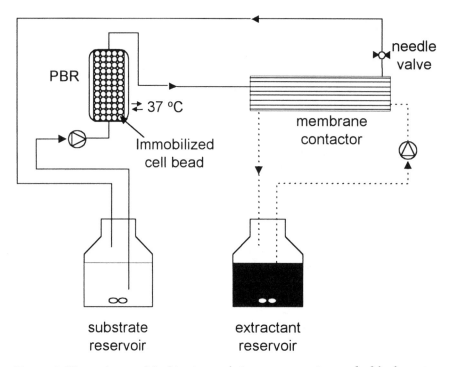

Figure 9. Flow scheme of the kinetic resolution process using packed-bed reactor.

The asymmetric synthesis of chiral amines with ω-transaminase has many potential advantages over kinetic resolution such as no need of reductive amination to prepare racemic amine from ketone precursor, and two times higher theoretical yield. However, as pointed out previously [8], the asymmetric synthesis reaction is very difficult to be carried out due to unfavorable thermodynamic equilibrium and much more severe product inhibition than the kinetic resolution reaction. For example, the theoretical maximal concentration of (S)-α-MBA through the asymmetric synthesis using acetophenone (30 mM) and L-alanine (300 mM) is only 2.7 mM. In addition to the thermodynamic limitation, (S)-α-MBA and pyruvate are much better substrates for the transamination reaction than acetophenone and L-alanine, suggesting that more severe product inhibition occurs in the amination of acetophenone than in the deamination of (S)-α-MBA. Therefore, at least one of the products should be removed during the reaction for the equilibrium shift and the reduction of product inhibition.

When L-alanine is used as an amino donor, the reaction yield can be increased dramatically by removing pyruvate, which is more inhibitory product. The removal of pyruvate was carried out by incorporating lactate dehydrogenase (LDH) in cell-free extract or by using whole cells. The whole cell reaction yielded a much better result. When 25 mM of benzylacetone and 30 mM of acetophenone were used as an amino acceptor with 300 mM of L-alanine, 90.2 and 92.1 % of the reaction yields were obtained after 1 day with whole cells, respectively. The enantiomeric excesses of both (S)-α-MBA and (S)-1-methyl-3-phenylpropylamine ((S)-MPPA) were all above 99 %.

Conclusions

Production of chiral amines (Figure 3) using ω-transaminase can be carried out via both kinetic resolution and asymmetric synthesis. The kinetic resolution would be better choice than the asymmetric synthesis due to much higher reaction rate and easy removal of the product inhibition. All the three processes for the kinetic resolution are feasible for scale-up. The choice of the production process depends on the economics of process accounting the availability of the enzyme and operating cost.

References

1. Juaristi, E.; Escalante, J.; Leon–Romo, J. L.; Reyes, A. *Tetrahedron: Asymmetry.* **1998**, *9*,715-740.
2. Kitaguchi, H.; Fitpatrick, P. A.; Huber, J. E.; Klibanov, A. M. *J. Am. Chem. Soc.* **1989**, *111*, 3094-3095.
3. Stirling, D. I. In *Chirality in industry*; Collins, A. N.; Sheldrake, G. N.; Crosby, J., Ed.; The use of aminotransferases for the production of chiral amino acids and amines; Wiley: New York, NY, 1992.

4. Gutman, A. L.; Meyer, E.; Kalerin, E.; Polyak, F.; Sterling, J. *Biotechnol. Bioeng.* **1992**, *40*; 760-767.
5. Stinson, S. C. *Chem. Eng. News.* **1996**, *July 15*, 35-61
6. Chapman, D. T.; Crout, D. H. G.; Mahmoudian, M.; Scopes, D. I. C.; Smith, P. W. *Chem. Commun.* **1996**, 2415-2416.
7. Shin, J.-S.; Kim, B.-G. *Biotechnol. Bioeng.* **1997**, *55*, 348-358
8. Shin, J.-S., Kim, B.-G. *Biotechnol. Bioeng.* **1998**, *60*, 534-540
9. Steinfeld, J. I.; Francisco, J. S.; Hase, W. L. *Chemical kinetics and dynamics;* Prentice Hall: Englewood Cliffs, 1989; pp 142-150
10. Kragl, U. In *Industrial enzymology;* Godfrey, T.; West, S., Ed.; Immobilized enzyme and membrane reactors; Stockton Press: New York, NY, 1996; pp 273-283.
11. Sheldon, R. A. *J. Chem. Tech. Biotechnol.* **1996**, *67*, 1-14
12. Christen, P.; Metzler, D. E. *Transaminases;* John Wiley & Sons Inc.: New York, NY, 1985.
13. Shin, J.-S., Kim, B.-G. *Biotechnol. Bioeng.* **1999**, *65*, 206-211
14. Shin, J.-S., Kim, B.-G. *Appl. Microbiol. Biotechnol.* **1999**, *submitted*

Chapter 17

Enantiorecognition of Chiral Acids by *Candida rugosa* Lipase: Two Substrate Binding Modes Evidenced in an Organic Medium

Per Berglund[1,3], Maria Christiernin[1], and Erik Hedenström[2]

[1]**Department of Biotechnology, Royal Institute of Technology, SE–100 44 Stockholm, Sweden**
[2]**Department of Chemistry and Process Technology, Mid Sweden University, SE–851 70 Sundsvall, Sweden**

We have identified the existence of different modes of binding the enantiomers of 2-methyl-branched carboxylic acids to a lipase active site by rational substrate engineering. Similar to hydrolysis, previously investigated, we have now evidence for differential binding modes in the *Candida rugosa* lipase-catalyzed esterifications in cyclohexane. The relevance of considering two different binding modes to understand lipase enantiorecognition is demonstrated by introducing bulky substituents on a chiral carboxylic acid which impose a different orientation of the substrate acyl chain in the active site of *Candida rugosa* lipase. With this substrate engineering approach based on molecular modeling it is thus possible to markedly alter the enantioselectivity of the lipase. Examples from hydrolysis and new results from esterifications in an organic solvent are presented and discussed.

Lipases are carboxyl ester hydrolases (E.C. 3.1.1.3) and accept a wide range of substrates. They are of frequent use in organic stereoselective synthesis (*1–5*). Many lipases with high stability are commercially available in pure and immobilized form. For a recent extensive review of lipase structure and applications, see Schmid and Verger (*6*). Today, stereoselective organic synthesis is facing explosive growth of new chiral unnatural compounds, which are of great interest in enantiomerically pure form in, for instance, the pharmaceutical and pesticide fields. Redesigning and controlling

[3]Corresponding author: email: Per.Berglund@biochem.kth.se.

lipase specificity for these new target molecules is a challenging task. Several strategies to alter and optimize lipase specificity for a certain unnatural synthetic target molecule have been explored and described in the literature, for instance directed evolution (7,8), and various gene-shuffling techniques (9,10). It is evident that creating an enzyme with tailor-made properties by site-directed mutagenesis alone requires more detailed knowledge of the mechanisms of lipase enantiorecognition.

Methyl-branched carboxylic acids of high enantiomeric purity are important building blocks in the synthesis of insect pheromones (11), chiral drugs (12), and liquid crystals (13). We have previously explored 2-methyl-branched chiral acids in lipase-catalyzed resolution reactions in organic media (14–16). For example, 2-methyloctanoic acid and 2-methyldecanoic acid have been prepared in >99.5% ee (16,17). A molecular modeling study of the esterification of 2-methyldecanoic acid with 1-heptanol has been undertaken which predicted the existence of two completely different modes of binding the enantiomers of such a chiral acyl donor to the active site of *Candida rugosa* lipase (18), Figure 1.

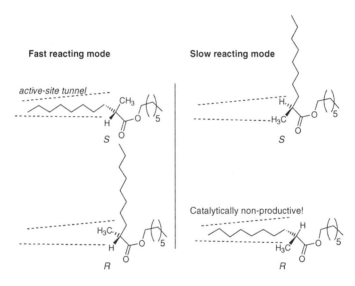

Figure 1. Predicted modes of binding the enantiomers of 1-heptyl 2-methyldecanoate in the active site of Candida rugosa lipase from molecular modeling (18).

The fast reacting S-enantiomer is suggested to orient its hydrophobic tail into the active-site tunnel in the normal mode, in accordance with the x-ray structure of a lipase-inhibitor complex (19). In contrast, in a similar mode of binding, the slow reacting R-enantiomer distorts the histidine 449 of the catalytic triad, which results in the loss of two catalytically essential hydrogen bonds to the transition state. The R-enantiomer must therefore bind to the enzyme active site leaving the tunnel empty,

Figure 1. Furthermore, this modeling study predicted a lower S-enantiomer preference and even a switch to R-enantiopreference in the case where both enantiomers were modeled orienting their hydrophobic tails out from the tunnel leaving the tunnel empty.

In order to explore the mechanisms of lipase enantiorecognition in more detail we have used a strategy where the fast-reacting S-enantiomer are forced to change its orientation from the normal mode of binding in the *C. rugosa* lipase to a situation orienting its hydrophobic tail out from the tunnel. This involved the design and synthesis of a diagnostic substrate. This substrate was sterically demanding but yet structurally analogous to 2-methyldecanoic acid and was expected to be too large to be accommodated in the active-site tunnel of the lipase active site. Both enantiomers would then bind in a similar way leaving the geometrically restricted active-site tunnel empty. In accordance with the predictions from previous molecular modeling studies (*18*), our diagnostic substrate ethyl 2-methyl-6-(2-thienyl)hexanoate gave a reversed enantiomer preference of the enzyme in the studied hydrolytic reactions (*20*), Scheme 1. This means that the existence of differential modes of binding enantiomeric substrates to the active site of *C. rugosa* lipase was confirmed in hydrolytic reactions.

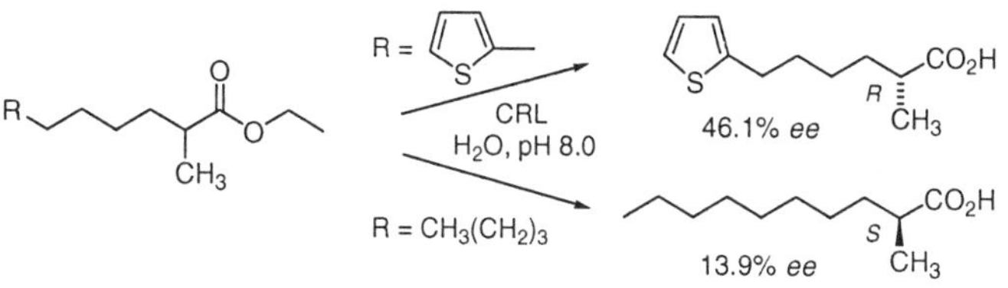

Scheme 1. Enantiomer preference in the Candida rugosa *lipase-catalyzed hydrolysis of 2-methylcarboxylic acid esters. Data are from reference 20.*

The hydrolytic reactions gave products with low enantiomeric excess. We have previously shown that the alcohol moiety of a 2-methyldecanoic acid ester influences the E-value of *C. rugosa* lipase and that ethyl esters give low E-values (*21*). Furthermore, we have shown that this lipase has a better performance in an organic solvent than in water when using 2-methylalkanoic acids or esters as substrates (*21*).

This suggests that with the same strategy higher E-values can be obtained in an organic solvent, and that a more dramatic change in the E-value would be seen due to the changed binding of the fast-reacting enantiomer in this case. In order to test this we have now explored the strategy in an esterification reaction in an organic solvent.

Results and Discussion

Synthesis

We synthesized three structurally analogous acids, 2-methyldecanoic acid, 2-methyl-6-(2-thienyl)hexanoic acid, and 2-methyl-3-(5-propyl-2-thienyl)propanoic acid and used them in the *Candida rugosa* lipase-catalyzed esterification in cyclohexane. The syntheses are detailed in Scheme 2.

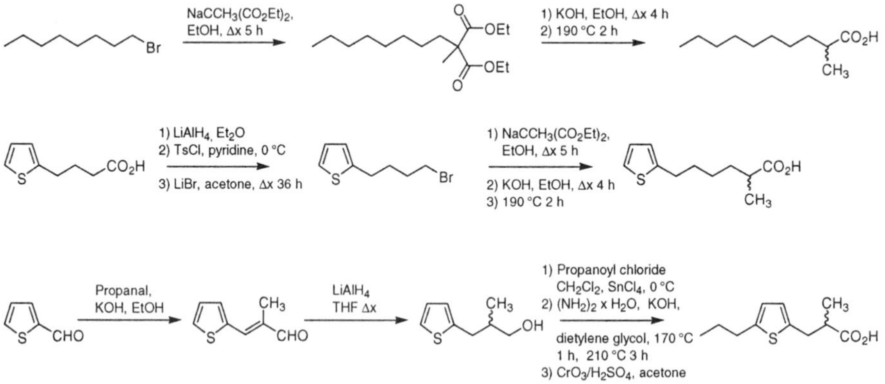

Scheme 2. Synthesis of the 2-methylcarboxylic acid substrates.

The advantage of using a substrate containing a thienyl group on the alkyl chain as an analog to the saturated four carbon unit is that the former one can easily be reduced to the saturated C_4 unit employing Raney-Ni conditions. The thienyl group is therefore an excellent candidate as a sterically demanding group since the saturated analog can be obtained with a different or even a reversed enantiomeric excess.

Esterification Reactions

The strategy of using the thienyl-substituted acid to evidence the existence of differential substrate binding modes is summarized in Figure 2.

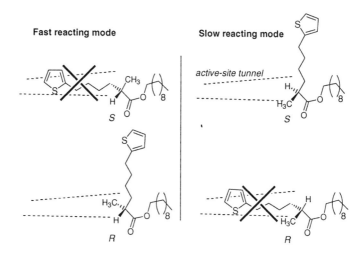

Fast reacting mode   **Slow reacting mode**

Figure 2. Summary of the strategy used. The bulky thienyl group forces the fast reacting S-enantiomer to adopt a conformation similar to the R-enantiomer leading to a situation of lower S-enantiomer preference of the enzyme.

The esterification reactions were run at room temperature in cyclohexane at a controlled water activity as recently described (*16*), Scheme 3. The progress of the reactions was followed by gas chromatography, and the extent of conversion was roughly calculated based on an internal standard. After terminating the reactions, product and remaining acid were isolated and their optical rotations were measured for the pure compounds. When traces of alcohol was present only the sign of the optical rotation is given.

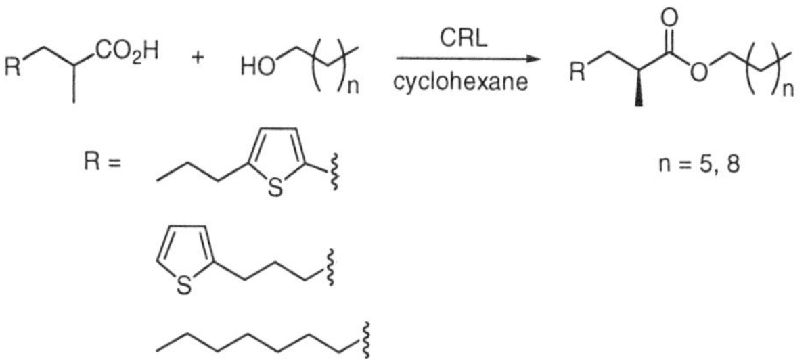

Scheme 3. The studied Candida rugosa *lipase-catalyzed esterification reactions of 2-methyl-branched carboxylic acids (150 mM) with primary alcohols (150 mM) in cyclohexane at a controlled water activity of* $a_w = 0.8$.

The product esters and remaining acids were derivatized to diastereomeric phenyl ethyl amides and the enantiomeric excess (ee) was indirectly calculated by the diastereomeric ratio obtained by GC or GC-MS. The E-value was calculated from the ee of the product (ee_{prod}) and of the substrate (ee_{sub}) using the equation derived by Rakels et al. (*24*). Data are presented in Tables I and II.

Table I. Enantioselectivity (E) and initial rates in the *C. rugosa* lipase-catalyzed esterfication reactions in Scheme 3.

Reaction		Conversion[a] (%)	ee_{sub} (%)	ee_{prod} (%)	E	Initial rate ($\mu mol\ min^{-1}\ g^{-1}$)
Acid, R	Alcohol, n					
~~{	5	23[b]	26.6[b]	96.3[b]	69[b]	6.5[b]
~~{	8	38	51.2	96.6	97	7.9
(thienyl)~~{	5	25	9.6	24.1	1.8	0.30[c]
(thienyl)~~{	8	36	16.0	21.0	1.8	0.44[c]
~~(thienyl){	8	37	5.5	9.7	1.3	0.22[c]

[a] Stages of conversion where the reactions were stopped and substrate and product isolated.
[b] Data from ref. *15*. [c] Based on two separate experiments.

Due to its bulkiness, a thienyl group on the acyl chain would restrict the fast reacting enantiomer to bind and react in the active-site tunnel and instead force it to bind in a hairpin fashion leaving the tunnel empty. These experiments were therefore expected to show a lower S-enantiomer preference compared to the experiment with the 2-methyldecanoic acid, without the thienyl group.

Table II. Specific optical rotation of the isolated substrate acids and product esters from the esterification reactions in Scheme 3.

Reaction		$[\alpha]_D^{20}{}_{sub}\ (CHCl_3)$	$[\alpha]_D^{20}{}_{prod}\ (CHCl_3)$
Acid, R	Alcohol, n		
~~{	8	No data	positive
(thienyl)~~{	5	No data	+2.71 (c 1.08)
(thienyl)~~{	8	−3.98 (c 1.78)	+2.45 (c 2.33)
~~(thienyl){	8	Negative	+1.11 (c 1.79)

E-values

Esterification reactions are reversible to some extent even in cyclohexane. In order to obtain accurate E-values from single point measurements using the equations derived by Sih, et al. (*22*), derived for irreversible reactions, the enantiomeric excess of the products have to be determined at a low conversion (*23*), preferably below 40%. Since the extent of conversion are often difficult to determine with an acceptable precision by GC, an alternative equation has been derived by Rakels et al (*24*), where E is calculated from the enantiomeric excess of both substrate and product. The enantiomeric excess of the remaining non-esterified acids and product esters were determined by conversion in several steps into diastereomeric N-1-phenylethyl amides since analysis of the enantiomeric excess of the nonracemic product esters is not possible by existing gas chromatographic or HPLC methods. Reduction by LiAlH$_4$ (25), as well as oxidation by Jones' reagent (26–28) are processes known to proceed without any racemization.

In accordance with the predictions of a lower S-enantiomer preference, the E-value changed from 97 to 1.8 on introducing the thienyl substituent at the 6-position of the acyl chain of the substrate. The optical rotation for the esters were low but still positive and for the remaining acid negative, indicating a small enantiomeric excess of S-ester product. When comparing the reactions of the 2-methyl-branched thiophene carboxylic acids with the 2-methyldecanoic acid, the former ones had an 18–36 fold lower reaction rate than the 7.9 μmol min^{-1} g^{-1} obtained for 2-methyldecanoic acid. It is thus possible that the thienyl group restricts the 2-methyl-branched thiophene carboxylic acids and that they possibly react with a higher fraction in the hairpin binding mode. When the thienyl moiety was moved from the end of the acyl chain to the middle, this lead to a slightly lower E-value of 1.3 and a reduced reaction rate with a factor of 0.5. Clearly, moving the slightly polar thienyl group closer to the reacting center somewhat decreases the rate of the S-enantiomer.

We have previously evidenced the existence of differential modes of binding substrate enantiomers in hydrolysis (20). The thiophene approach then lowered the S-enantiomer preference from E = 1.4 to an R-enantiopreference of E = 3.6. This corresponds to a change in the free energy difference between the diastereomeric transition states of $\Delta\Delta\Delta G^{\neq} = 3.9$ kJ mol^{-1}. In this work we have lowered the S-enantiomer preference from E = 97 to E = 1.3, corresponding to a free energy change of $\Delta\Delta\Delta G^{\neq} = 10.5$ kJ mol^{-1}. Clearly, using an organic solvent system and a longer alcohol than in the case of hydrolysis dramatically amplifies the effect of the thienyl group. This is in accordance with previous studies, where it has been shown that longer alcohols lead to higher E-values in *C. rugosa* lipase-catalyzed esterifications of 2-methyldecanoic acid (14,17). Hydrolysis of a 2-methyldecanoic acid ester of a long alcohol, such as octanol, leads to a low equilibrium position and the reaction appears to stop around 15% conversion (21). This fact makes it difficult to study the effect of a longer ester in hydrolysis.

It is evident that we have accomplished a more dramatic change of the E-value in this work, which clearly support the evidence for the existence of multiple modes of binding enantiomeric alkanoic acids to the active site of *Candida rugosa* lipase.

Conclusions

We have shown that it is possible to tailor a substrate molecule by engineering the steric restrictions to force it to adopt a different orientation of its acyl chain in the enzyme active site leading to a dramatically altered enantioselectivity of the lipase. With a substrate engineering approach it is thus possible to markedly alter the enantioselectivity of *Candida rugosa* lipase for a chiral carboxylic acid in esterifications. These data, together with the previously published data on hydrolysis (*20*), constitute kinetic evidence for the existence of different modes of binding enantiomeric substrates into an enzyme active site and contribute to the understanding of the mechanisms of enzyme enantiorecognition. The relevance of considering multiple modes of binding enantiomers to an enzyme is specifically important in the modeling of unnatural substrates of interest for synthetic chemists.

Experimental Section

Materials

Candida rugosa lipase (EC 3.1.1.3) Sigma lipase type Vll, with the specific activity stated as 950 units per mg solid 4,540 units per mg protein were obtained from Sigma, St Louis, MO, USA. Macroporous polypropylene powder, Accurel EP 100, 350–1000 μm was obtained from Akzo Faser AG, Obernburg, Germany. Octadecane was from Sigma, St Louis, MO, USA. 1-Decanol and 1-heptanol were purchased from Lab-Scan, Dublin, Ireland. (*R*)- and (*S*)-1-phenylethylamine, >99.8% ee, was from Fluka Chemicals AG, Switzerland. Powdered LiAlH$_4$, >98% purity, was obtained from Kebo lab, Sweden.

Substrate Acids

The carboxylic acid substrates 2-methyldecanoic acid, 2-methyl-6-(2-thienyl)hexanoic acid, and 2-methyl-3-(5-propyl-2-thienyl)propanoic acid were synthesized as detailed in Scheme 2 in good total yields and in >99.8%, >99%, and 98.5% purity, respectively (*29*).

Immobilization of the Lipase

Crude lipase (200 mg) was dissolved in sodium phosphate buffer (20 ml, pH 7, 20 mM) and the solution was centrifuged for 2 min at 800 × g. Ethanol (1 ml) was added to the polypropylene carrier (100 mg) and the mixture was degassed under vacuum. The supernatant was then mixed with the carrier and the suspension was allowed to rotate in an end-over-end rotation device for 22 h at room temperature. After filtration and drying under vacuum the immobilized lipase was pre-equilibrated to a water activity of $a_w = 0.76$ above a saturated aqueous NaCl solution in a closed vessel for 24 h.

Esterification Reaction

2-Methylcarboxylic acid (0.15 M), the alcohol (0.15 M), and octadecane used as internal standard (12.2 mg, 48 mmol) were dissolved in cyclohexane (2 ml). The water activity (a_w) was kept constant at 0.8 throughout the reactions by adding solid anhydrous Na_2SO_4 (56.8 mg, 0.4 mmol) and $Na_2SO_4 \times 10H_2O$ (64.4 mg, 0.2 mmol). Immobilized pre-equilibrated enzyme (70 mg) was added to the mixture, which was left to react in an end over end rotation device at room temperature.

Determination of Conversion

To follow the reactions samples of approximately 10 μl were taken from the reaction mixtures at various intervals and were diluted with cyclohexane (90 μl). The initial reaction rates were calculated for conversions below 15% since all reactions showed linearity below this point. The reactions were stopped close to 40% conversion by filtering off the enzyme. The conversion was followed by a Perkin Elmer 8420 capillary gas chromatograph fitted with fused silica 25 m × 0.32 mm ID columns coated with either Cpwax 58CB, df 0.2 μm, or Cpsil 5CB, df 0.2 μm, both from Chrompac. Oven temperatures was programmed from 240–270 °C and nitrogen or helium (6.5 kPa, 1.5 ml/min) was used as carrier gas.

Separation of Compounds

Ester, alcohol, acid, and internal standard were separated by medium performance liquid chromatography (MPLC) on a column with silica gel 60 (10 g, 230–400 mesh). Compounds were eluted starting with pure cyclohexane and then an increasing gradient of ethyl acetate up to 30%. In the second to last fraction pure ethyl acetate was used and in the last 10% acetic acid in ethyl acetate was employed. TLC and GC were used to retrieve the fractions of interest.

Determination of Enantiomeric Excess

To measure the enantiomeric excess of the product (ee_{prod}) the esters were reduced to alcohols, then oxidized to acids and finally derivatized to diastereomeric phenylethyl amides (14,16). The glassware used for reduction was dried in an oven and allowed to cool in a desiccator. Ester (0.054 mmol) was dissolved in anhydrous diethyl ether

(1 ml). In a 10 ml vessel with anhydrous diethyl ether (1 ml) under nitrogen 4 times molar excess of powdered LiAlH$_4$ (10 mg app. 0.2 mmol) was added. The ester solution was added to the LiAlH$_4$ solution with a syringe. The mixture was then left to react for 1 h under nitrogen and stirring. The reaction was then quenched with water (20 μl), 15% NaOH (20 μl), and water (40 μl). The mixture was finally refluxed for 1 h and then filtered and dried with MgSO$_4$. The solvent was evaporated and the resulting alcohols were dissolved in acetone (1 ml). Jones' reagent [2.67 M, prepared by dissolving CrO$_3$ (26.72 g) in water (100 ml) and then adding conc. H$_2$SO$_4$ (23 ml)] (30) in 3 times volumetric excess was added and the temperature was kept at 5 °C. After a few minutes celite was added and the mixture was filtered. Some of the acids were extracted with 10% Na$_2$CO$_3$, followed by acidification before amides were synthesized yielding less but purer product. Others were derivatized directly without further purification. Amidification process: Acid, (5 μl) was dissolved in diethyl ether (1 ml), dry dimethyl formamide (5 μl) and thionyl chloride (5 μl) were added under nitrogen. Enantiomerically pure N-1-phenylethyl amine (10 μl) was added, (R)-(+)-N-1-phenylethyl amine was used for all products from the enzymatic reactions and (S)-(-)-N-1-phenylethyl amine for the substrate acids. Nitrogen was used to flush away the HCl formed. The ether phase was washed with water (1 ml) which dissolved the precipitate. After vigorous shaking the ether phase was washed twice with 10% Na$_2$CO$_3$ (2 × 1 ml) and once with saturated NaCl solution (1 ml). The ether phase containing the diastereomeric mixture of the phenylethyl amides was then analyzed without further purification with GC or GC-MS and ee$_{prod}$ and ee$_{sub}$ calculated.

Acknowledgments

Financial support from the Swedish Council for Forestry and Agricultural Research (SJFR) and from the Swedish Natural Science Research Council (NFR) is gratefully acknowledged.

Literature Cited

1. Wong, C.-H.; Whitesides, G. M. *Enzymes in Synthetic Organic Chemistry;* Tetrahedron Organic Chemistry Series 12; Pergamon: Oxford, U.K., 1994.

2. Faber, K. *Biotransformations in Organic Chemistry,* 3rd ed.; Springer: Berlin, Germany, 1997.

3. *Lipases part B;* Rubin, B.; Dennis, E. A., Eds.; Methods in Enzymology; Academic Press: New York, 1997; Vol. 286, pp1–563.

4. Kazlauskas, R. J.; Bornscheuer, U. T. In *Biotransformations I;* Kelly, D. R., Ed.; Biotechnology, 2nd ed.; Wiley: Weinheim, Germany, 1998; Vol. 8a, pp 37–191.

5. Berglund, P.; Hult, K. In *Stereoselective Biocatalysis;* Patel, R., Ed.; Marcel Dekker: New York, 1999; pp 633–657, in press.

6. Schmid, R. D.; Verger, R. *Angew. Chem. Int. Ed. Engl.* **1998**, *37*, 1609–1633.

7. Reetz, M. T.; Zonta, A.; Schimossek, K.; Liebeton, K.; Jaeger, K.-H. *Angew. Chem. Int. Ed. Engl.* **1997**, *36*, 2830–2832.

8. Arnold, F. H.; Volkov, A. A. *Curr. Opin. Chem. Biol.* **1999**, *3*, 54–59.

9. Nixon, A. E.; Ostermeier, M.; Benkovic, S. B. *Trends Biotechnol.* **1998**, *16*, 258–264.

10. Holmquist, M.; Berglund, P. *Org. Lett.* **1999**, *1*, 763–765.

11. Högberg, H.-E.; Berglund, P.; Edlund, H.; Fägerhag, J.; Hedenström, E.; Lundh, M.; Nordin, O.; Servi, S.; Vörde. C. *Catalysis Today* **1994**, *22*, 591–606.

12. Sheldon, R. A. *Chirotechnology: Industrial Synthesis of Optically Active Compounds*; Marcel Dekker: New York, 1993.

13. Bydén, M.; Edlund, H.; Berglund, P.; Lindström, B. *Prog. Colloid. Polym. Sci.* **1997**, *105*, 360–364.

14. Berglund, P.; Holmquist, M.; Hedenström, E.; Hult, K.; Högberg, H.-E. *Tetrahedron: Asymmetry* **1993**, *4*, 1869–1878.

15. Berglund, P.; Holmquist, M.; Hult, K.; Högberg, H.-E. *Biotechnol. Lett.* **1995**, *17*, 55–60.

16. Berglund, P.; Hedenström, E. In *Enzymes in non-aqueous media: IV Synthetic aspects;* Holland, H., Ed.; Methods in Biotechnology; Humana Press: Totowa, NJ, 1999; *In Press*.

17. Edlund, H.; Berglund, P.; Jensen, M.; Hedenström, E.; Högberg, H.-E. *Acta Chem. Scand.* **1996**, *50*, 666–671.

18. Holmquist, M.; Hæffner, F.; Norin, T.; Hult, K. *Protein Sci.* **1996**, *5*, 83–88.

19. Grochulski, P.; Bouthillier, F.; Kazlauskas, R. J.; Serreqi, A. N.; Schrag, J. D.; Ziomek, E.; Cygler, M. *Biochemistry* **1994**, *33*, 3494–3500.

20. Berglund, P.; Holmquist, M.; Hult, K. *J. Mol. Catal. B: Enzym.* **1998**, *5*, 283–287.

21. Holmberg, E.; Holmquist, M.; Hedenström, E.; Berglund, P.; Norin, T.; Högberg, H.-E.; Hult, K. *Appl. Microbiol. Biotechnol.* **1991**, *35*, 572–578.

22. Chen, C.-S.; Fujimoto, Y.; Girdaukas, G.; Sih, C. J. *J. Am. Chem. Soc.* **1982**, *104*, 7294–7299.

23. Van Tol, J. B. A.; Jongejan, J. A.; Geerlof, A.; Duine, J. A. *Recl. Trav. Chim. Pays-Bas* **1991**, *110*, 255–262.

24. Rakels, J. L. L.; Straathof, A. J. J.; Heijnen, J. J. *Enzyme Microb. Technol.* **1993**, *15*, 1051–1056.

25. Noyce, D. S.; Denney, D. B. *J. Am. Chem. Soc.* **1950**, *72*, 5743–5745.

26. Sonnet, P. E. *J. Org. Chem.* **1982**, *47*, 3793–3796.

27. Sonnet, P. E. *J. Org. Chem.* **1987**, *52*, 3477–3479.

28. Guanti, G.; Narisano, E.; Podgorski, T.; Thea, S.; Williams, A. *Tetrahedron* **1990**, *46*, 7081–7092.

29. Hedenström, E.; Department of Chemistry and Process Technology, Mid Sweden University, Sundsvall, Sweden; Unpublished Results.

30. Fieser, L. F.; Fieser, M. *Reagents for organic synthesis*; Wiley: New York, 1967; Vol. 1, p 142.

Author Index

Subject Index

activity of various immobilized PS-30
preparations, 160
bioreactor design, 157–158
effect of alkylammonium salts on immobilized PS-30 catalyzed esterification of lauric acid with 1-octanol,
161f
enzymatic activity of immobilized lipoxygenase (LOX), 158, 159f
immobilization conditions, 158, 160
kinetics of esterification of lauric acid
with 1-octanol catalyzed by immobilized PS-30, 161f
lipoxygenase bioreactor design, 159f
lipoxygenase bioreactor for continuous
oxygenation of linoleic acid, 162
materials, 156
materials and methods, 156–158
measurement of lipase PS-30 activity
method, 157
measurement of lipoxygenase activity
method, 157
method for entrapment of enzymes,
157
novel procedure for intercalative immobilization of LOX within, 156
producing hydroperoxy fatty acid
(HPOD) using immobilized lipoxygenase, 156
recycling activity of free PS-30 and
phyllosilicate sol-gel immobilized
PS-30, 163f
reusability of immobilized PS-30, 160
Pig pancreatic lipase, organic solvents,
29
Polarizing field, radiation theory for enzyme activity, 16
Polymerase chain reaction (PCR)
overlapping PCR for gene shuffling,
78
See also Gene shuffling
Porphyrins, replacing iron porphyrin
with manganese porphyrin in horseradish peroxidase, 28–29
Precipitable indigogenic substrates, class
of substrates, 6
Preparative chemistry, economics of coenzymes, 105, 106f
Proteases, detergents, 7
Proteins, notion that enzymes are, 16–17
Publishing, establishing scientific priority, 33–34

Pyridine coenzymes
relation of functional groups to function, 103
See also Nicotinamide cofactors
Pyridine nucleotides
electrochemical regeneration, 119
role in biotechnology industry, 105
Pyridylacetates, hydrolysis
hydrolysis of pyridylacetates of secondary alcohols, 212t
kinetic resolution of pyridylacetates using penicillin acylase, 205t
penicillin acylase, 204t
See also Penicillin acylase catalyzed ester hydrolysis

R

Radiation theory, enzyme activity, 16
Random mutagenesis, gene shuffling of
aminopeptidase in combination with,
86, 88
Reaction processes, chiral amines
comparison of reaction processes, 259
enzyme membrane reactor, 257, 258f
packed-bed reactor (PBR), 259, 260f
two-liquid-phase reaction system, 257
Reduction, stereoselective. See Aminoalcohols, chiral vicinal
Regeneration
cofactor recycling for enzymatic conversion of sugars to molecular hydrogen, 125
enzymatic methods for, of cofactors,
108–110
NAD(P)$^+$, 110
NAD(P)H, 108, 110
primary methods for cofactor, 107
reduced or oxidized cofactor, 105, 106f
α-Rhamnosidase, fruit juice debittering,
9
Rhodococcus rhodochrous, nitrile hydratase converting acrylonitrile to acrylamide, 8
Ricinoleic acid
castor oil derivative, 92–93
conversion to 7,10,12-trihydroxy-8(E)-
octadecenoic acid (TOD), 96, 99
degradation and bioconversion products, 61
isolation of intermediate in production
of TOD, 98f
See also Unsaturated fatty acids

S

Saprophytic bacteria selection
  compost source materials, 56
  conversion of oleic acid by *Sphingo-
    bacterium thalpophilum* strains iso-
    lated from compost cultures, 61*t*
  degradation and bioconversion prod-
    ucts from linoleic acid, 62
  degradation and bioconversion prod-
    ucts from oleic acid and 10-ketos-
    tearic acid (KSA), 60–61
  degradation and bioconversion prod-
    ucts from ricinoleic acid (12-hy-
    droxy-9-octadecenoic acid), 61
  degradation and oxidation of long
    chain unsaturated fatty acids
    (UFAs), 56
  degraded or bioconverted products
    from specific fatty acids, 59*t*
  enrichment culture procedure, 56–57
  methods for identification of bacteria
    and their reaction products, 57
  typical gas chromatograms of biocon-
    version products, 58*f*
Screening hydrolase libraries
  actual substrate assays, 45–46
  assay, 50–52
  basic principles supporting pH-shift re-
    agent assay, 47*f*
  best library substrates for Amano li-
    pase PS, 50*t*
  color gradient of phosphate buffer con-
    taining BTB at different pH values,
    51, 54*a*
  enantioselectivity of hydrolases, 46
  enantioselectivity relation to time for
    color change of enantiomers, 51
  enzyme library, 49–50
  Kazlauskas Quick E assay for apolar
    substrates of library, 52*f*
  pH-responsive method for rapid
    screening, 45–46
  pH-shift reagents, 47
  results of 384-well format enantioselec-
    tivity assay of substrates, 51, 54*a*
  study with Bromothymol Blue (BTB),
    47
  substrate library, 48
  substrate library for hydrolase enantio-
    selectivity assay, 49*f*
  three different concentrations of phos-
    phate buffer containing BTB and ef-

  fect of increasing concentrations of
    acid, 51, 54*a*
  typical assay conditions, 50–51
  use of enzymes at off-neutral pH val-
    ues, 47
Screening strategies
  brute force method, 5–6
  hierarchical method, 5–6
  substrates and techniques, 6
  See also Enzyme screening
Sebacic acid, castor oil derivative, 92–93
Separation, oligosaccharides, 147–148
Serendipity, lysozyme, 34
Soil bioremediation
  degradation and oxidation of long
    chain unsaturated fatty acids
    (UFAs), 56
  See also Saprophytic bacteria selection
Sol-gel matrix
  immobilization of enzymes, 155–156
  See also Phyllosilicate sol-gel immobi-
    lized enzymes
Specificity
  enzyme, historical, 30–32
  thoughts on adsorption, activation,
    and specificity of active site, 16
  ω-transaminase from *V. fluvialis*, 250
Spiritus Vitae, ordinary physical force
  versus organic vital force, 15
Stability, coenzymes, 107–108
Starch
  adsorption of amylase, 28
  industrial applications of enzymes, 25*t*
Stemphylium loti, waste treatment, 11
Stereoselective reduction
  experimental conditions, 197–198
  See also Aminoalcohols, chiral vicinal
Steroid functionalization, hydroxylation and dehydrogenation, 124–125
Sugars, enzymatic conversion to molec-
  ular hydrogen, 125, 126*f*

T

Tanning industry, historical enzyme ap-
  plication, 22
Temperature, effect on fat unsaturation,
  35
Tenderizer, meat, papain, 26–27
Textile, industrial applications of en-
  zymes, 25*t*
Thermolysin, aspartame, 9
Timed-release formulation, Enzypan, 27